# THE SHORT STORY OF SCIENCE

First published in Great Britain in 2022 by
Laurence King Publishing Ltd
Carmelite House
50 Victoria Embankment
London EC4Y 0DZ

An Hachette UK Company

1 3 5 7 9 10 8 6 4 2



A CIP catalogue record for this book is available from the British Library.

ISBN 978-1-91394-788-0

John Round Design
Origination by DL Imaging, London
Printed in China by Asia Pacific Offset Ltd

**p.10** Persian scientist-philosopher al-Razi/Wellcome Collection
**p.41** Foucault's pendulum in the Panthéon, Paris/Rémih
**p.154** The Andromeda Galaxy/NASA/JPL/California Institute of Technology
**p.180** Installing the ATLAS detector at the Large Hadron Collider, CERN, Switzerland/CERN

Laurence King Publishing is committed to ethical and sustainable production. We are proud participants in The Book Chain Project bookchainproject.com®

www.laurenceking.com
www.orionbooks.co.uk

# THE SHORT STORY OF SCIENCE

## A Pocket Guide to Key Histories, Experiments, Theories, Methods & Equipment

**Tom Jackson**

Laurence King Publishing

# Contents

**THEORIES**



**METHODS AND EQUIPMENT**

# Introduction

**STEPHEN HAWKING:** 'WE ARE JUST AN ADVANCED BREED OF MONKEYS ON A MINOR PLANET OF A VERY AVERAGE STAR. BUT WE CAN UNDERSTAND THE UNIVERSE. THAT MAKES US SOMETHING VERY SPECIAL.'

Science is a means of revealing facts that have always existed, but which were previously unknown. Scientists probe the nature of material and strain to see to the very edge of the Universe – and perhaps beyond.

This book tracks the breakthroughs that underpin our current understanding of the Universe. It covers physics, which seeks out the laws that govern energy, matter and motion; chemistry, which investigates substances of all kinds and seeks to understand how one substance can transform into another; and biology, which studies forms of life. Additionally, it touches on psychology, astronomy, neuroscience and geology.

Science requires that new discoveries are built on older ones. The short stories recounted here throw light on scientific understanding and show how that understanding has grown steadily over centuries of work.

## Histories

**MARIE CURIE:** 'NOTHING IN LIFE IS TO BE FEARED, IT IS ONLY TO BE UNDERSTOOD. NOW IS THE TIME TO UNDERSTAND MORE, SO THAT WE MAY FEAR LESS.'

It took thousands of years for science to emerge from the fog of superstition and dogma. Today's scientists use rigorous methods that guide thinking, test ideas and carefully review results. However, the simple intuition that truth can be found by seeking evidence for it is a strong one that reaches back to the ancient world.

Science became the dominant force it is today only around 350 years ago, at the start of the period now known as the Scientific Revolution, during which the broad field of study gradually professionalized into specialist fields. By the 1850s, scientists had become experts with diverse interests such as cell biology, the nature of electricity, or the weight of atoms; it was no longer possible for any one person to be an expert in all the sciences.

## Experiments

**CARL SAGAN:** 'SOMEWHERE, SOMETHING INCREDIBLE IS WAITING TO BE KNOWN.'

The scientific process involves many steps but the most iconic is the experiment, in which a theory is tested and shown to be true or false. Experiments need not be elaborate to have an impact. In the 1950s, Stanley Miller created a primeval soup in a round flask and some glass tubes. Within a week his set-up had spontaneously created the chemicals needed for life.

## Theories

**ALBERT EINSTEIN:** 'TWO THINGS ARE INFINITE: THE UNIVERSE AND HUMAN STUPIDITY; AND I'M NOT SURE ABOUT THE UNIVERSE.'

Science is a creative process. We can only unlock hidden truths once we have imagined them. These imaginings are known as theories or hypotheses. Once proven – if proven – then a theory is no longer a theory; it is established and accepted as fact, although, perhaps confusingly, some theories continue to be termed theories even after they have been proved correct.

## Methods and Equipment

**ISAAC NEWTON:** 'IF I HAVE SEEN FURTHER IT IS BY STANDING ON THE SHOULDERS OF GIANTS.'

The history of science runs hand in hand with the history of technology. New science provides the insights required to develop new technology, and that technology in turn offers scientists new ways of investigating. There have been several great leaps forward in the instruments and technical aids available to science. In the sixteenth century CE, lens makers developed telescopes and microscopes and glass tubing that was so sensitive to changes in external temperature that it could be used in the construction of thermometers. Improvements in metal-refining made it possible to construct precision mechanisms and electrical and magnetic devices. Today's science relies on computing to control instruments and collect and analyze data, and it is hard to overestimate the extent to which computers have enhanced the practice of science. In all likelihood, new forms of computer technology will have a profound impact on the future of research.

## A Note on Context

Most advances in science have leapt from springboards of collective effort: few pioneers have worked alone, and even solitary scientists base their findings on pre-existent data.

While increasing public awareness of science is a laudable aim – indeed, it is the principal purpose of the present volume – one of its potential shortcomings is oversimplification: a tendency to attribute major achievements to a few, named individuals, even when the triumphs have been the work of vast teams, often cooperating or competing with each other in widely spread locations across the world.

Most of those thus acclaimed have been white males.

This cannot be shrugged off as journalistic shorthand: it is worse than inaccurate; it is unfair, and it damages people of colour and women by denying them due credit and by perpetuating the notion that they cannot possibly succeed in these fields.

Today there are increasing efforts to redress this imbalance, to give previously underacknowledged or uncredited people the recognition they deserve. It is noteworthy, however, that in the West scientists of African and Asian heritage were excluded for so long that contemporary coverage tends to be of the first minority people to come to the fore in various disciplines. Clearly, there is still much to be done to achieve equality of opportunity for non-white people.

Women have long made significant contributions to the fund of scientific knowledge, even during the periods when they were excluded from higher education and academia, but shamefully the credit for their work was for centuries routinely claimed by or given to men. Women's contributions have previously been downplayed or airbrushed out of the general picture, so it is hoped that readers of this book will note in particular that the means to measure the vastness of the Universe was the work of a woman (page 28); that a woman revealed the process of nuclear fission within atomic bombs (page 136); that the existence of dark matter, one of the biggest science mysteries, was proven by a woman (page 175); and that the Greenhouse Effect was first shown by a woman (page 178).

Science is not for the few; it is – or at least it should be – for everyone, always.

# How to use this book

This book is divided into four sections: Histories; Experiments; Theories; and Methods and Equipment. Each section can be read in isolation or in conjunction with other sections. Useful cross-references at the bottom of each page guide the reader from one section to another, while feature boxes provide biographical details about key scientists and a wealth of additional information.

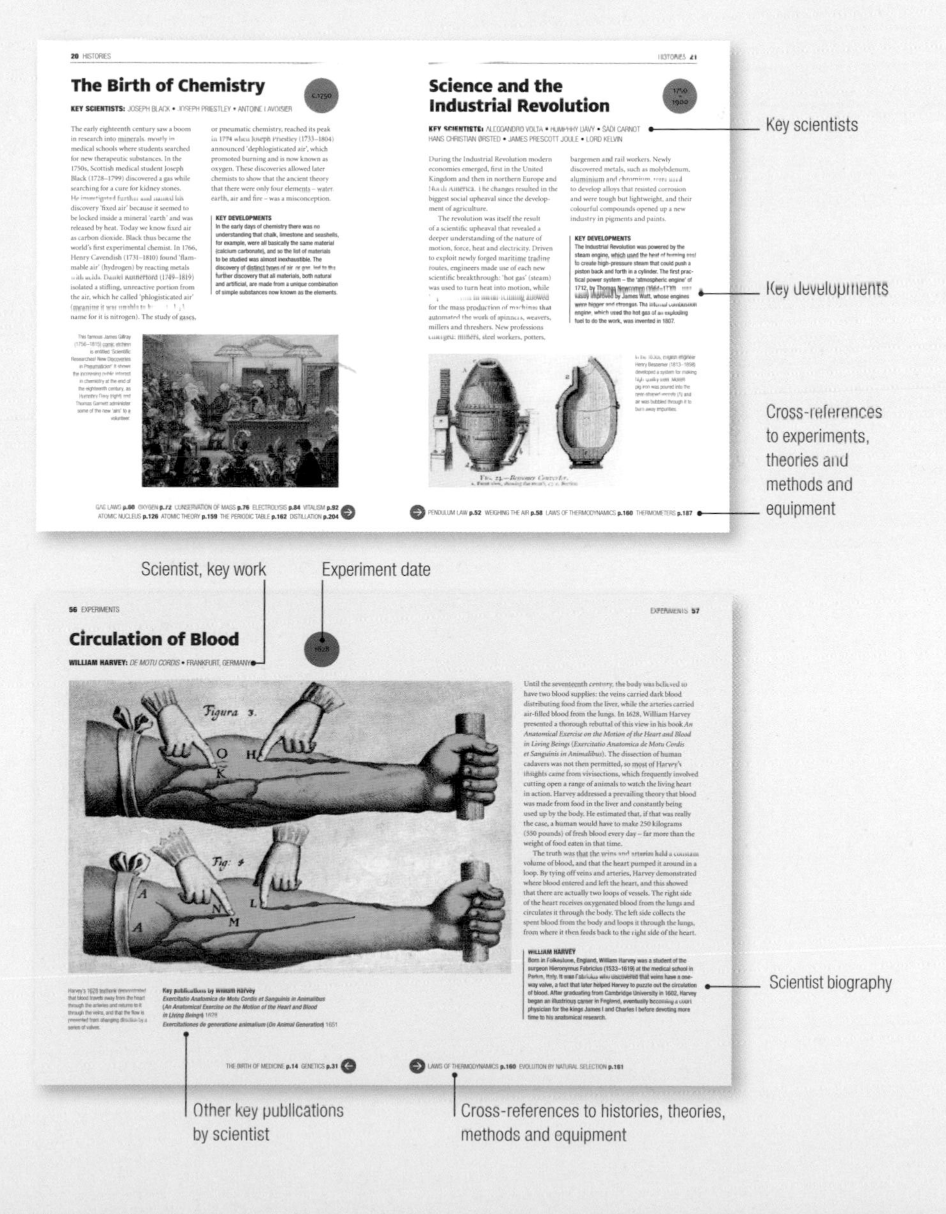

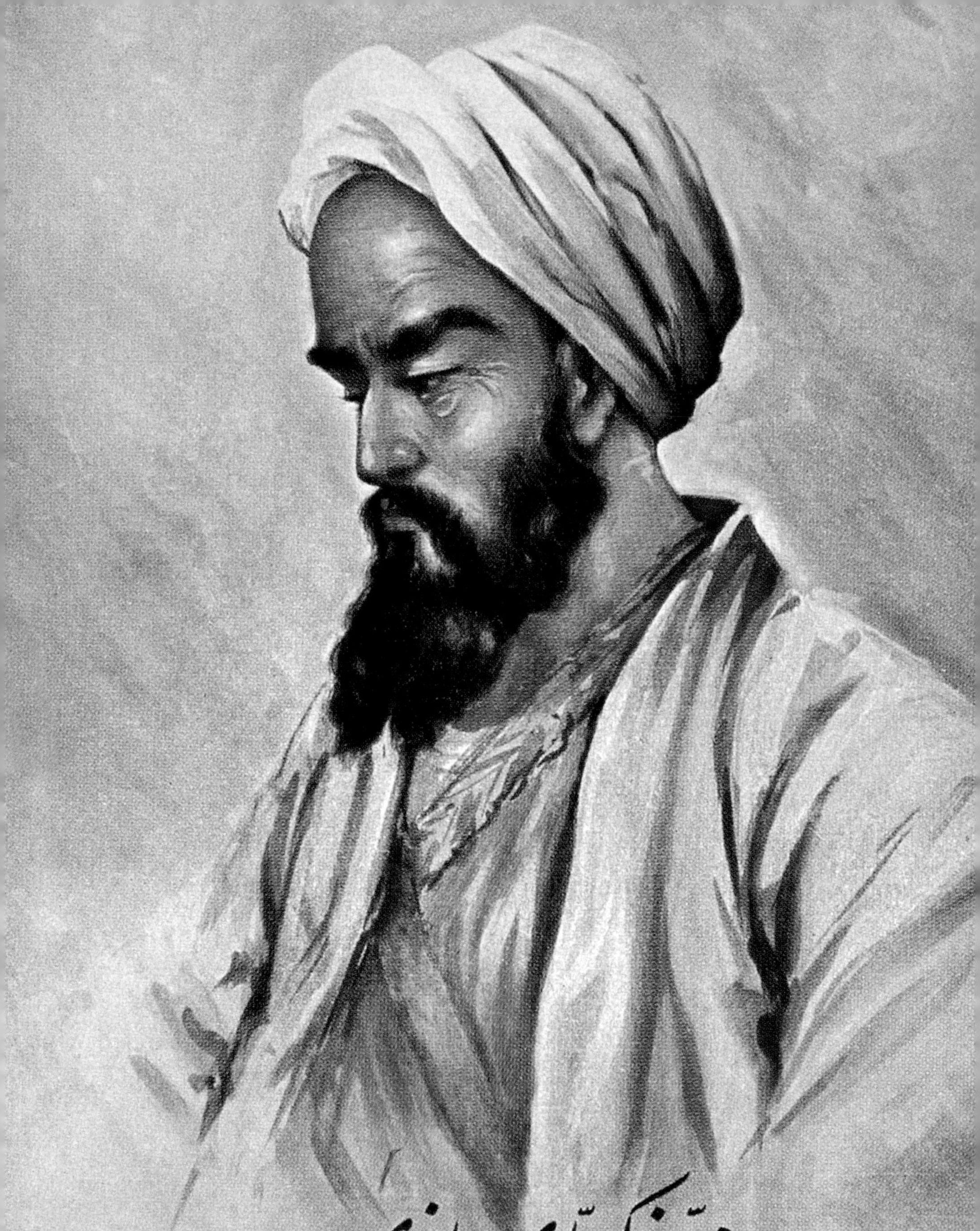

محمد زکریای رازی

(۲۵۰ - ۳۱۳ هجری)

هدیه سرویس طبی تجارتخانه حسن خسروشاهی بمناسبت پنجمین کنگره پزشکی رامسر ضمیمه نشریه «در کلینیک چه میکنند»

... دکتر محمود نجم آبادی و لوئی فنک مهرماه ۱۳۳۵

# HISTORIES

# Ancient Astronomers

**KEY SCIENTISTS:** EUDOXUS • ARISTOTLE

**KEY DEVELOPMENTS**

Ancient astronomers did exactly what modern ones do: they built models of the Universe based on what they could see of it. By the fourth century BCE, the dominant model placed Earth at the centre with the Moon, Sun and planets moving around it in perfect circles. The stars were set in the outermost sphere. This view of our place in the Universe was widely accepted for thousands of years.

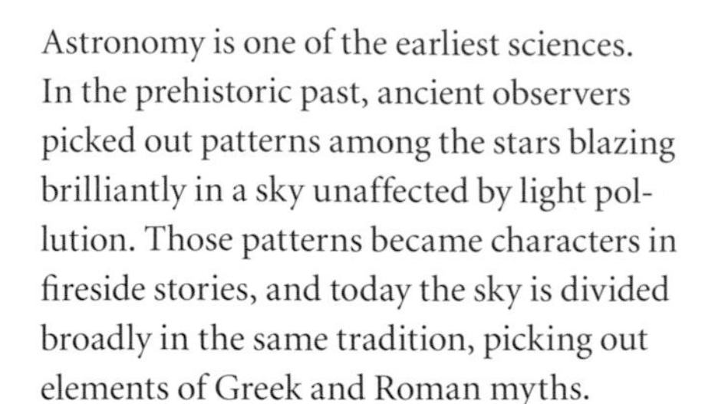

Astronomy is one of the earliest sciences. In the prehistoric past, ancient observers picked out patterns among the stars blazing brilliantly in a sky unaffected by light pollution. Those patterns became characters in fireside stories, and today the sky is divided broadly in the same tradition, picking out elements of Greek and Roman myths.

However, there was more to stargazing as people watched the constellations move hour by hour, night by night and month by month. They matched these movements to the rhythms of sunrise and sunset and the changing aspects of the Moon. They then used their observations to create calendars to mark out the changes of the seasons, so crucial for ensuring a successful harvest and maintaining food supplies throughout the year. Five heavenly bodies looked like stars but moved along their own path. These were the 'wanderers', or *planeta* in Latin, all of which are seen to move within a narrow band of the sky, which is still known by the Greek-derived name 'zodiac'.

Among the earliest known astronomers were the Babylonians, who linked the main celestial objects to their deities. This monumental stone of the twelfth century BCE depicts a star (the goddess Ishtar), the Moon (the god Sin) and the Sun (the god Shamash).

THE SIZE OF EARTH **p.44** EXOPLANETS **p.148** PANSPERMIA **p.156**
ORIGIN OF THE SOLAR SYSTEM **p.179** TELESCOPES **p.189**

# Greek Philosophers

**KEY SCIENTISTS:** PLATO • ARISTOTLE • THALES

*The School of Athens* (1509–1511) by Raphael (1483–1520) depicts Plato, Aristotle and other leading philosophers of ancient Greece.

There are a few contenders for the title of 'first scientist', including Imhotep (active c.2667–2648 BCE), architect of Egypt's first pyramid, and Sushruta (active c.600 BCE), a doctor working in India. However, the honour is usually bestowed on Thales of Miletus (c.624–c.548 BCE), a pioneer of natural philosophy. He and those who followed him sought to explain the world without recourse to mythology or religion. This tradition probably arose in ancient Greece rather than elsewhere because the Olympian pantheon of gods behaved in very human and irreligious ways and thus offered few compelling answers to the big questions. Nevertheless, Thales's methods of reasoning were nothing like what we regard as scientific today. In the fifth century BCE, Athens became the centre of natural philosophy, first with Socrates (c.470–399 BCE) teaching Plato (c.424–c.348 BCE), who mentored Aristotle (384–322 BCE). The last two disagreed about the source of knowledge. Plato believed that it was to be found only in supernatural ideals, while Aristotle insisted that the truth was revealed by examining what could be observed around us.

**KEY DEVELOPMENTS**

Among Greek philosophy's lasting legacies was the codification of logic. Deduction uses a logical consequence of two premises, for example: all men are mortal; Plato is a man; therefore, Plato is mortal. Induction is based on a weaker link between premise and consequence: the Sun rises every morning; therefore, the Sun will rise tomorrow. Both forms have been picked apart many times by later philosophers, but they are still crucial tools in the scientific process.

→ BUOYANCY p.42 THE SIZE OF EARTH p.44 PANSPERMIA p.156

# The Birth of Medicine

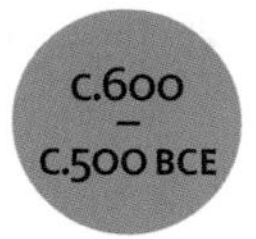


**KEY SCIENTISTS:** HIPPOCRATES • GALEN • AVICENNA

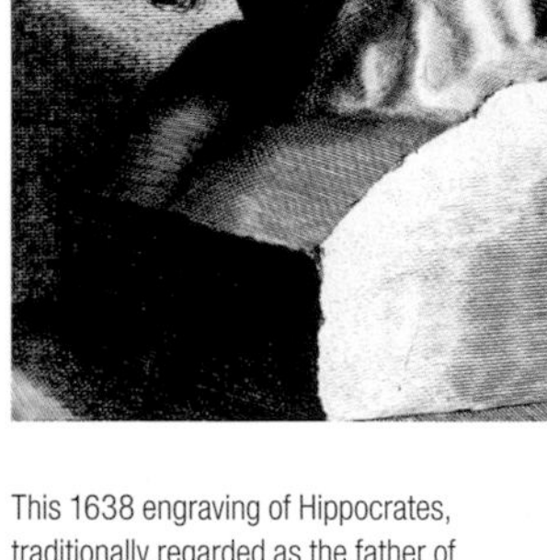

This 1638 engraving of Hippocrates, traditionally regarded as the father of medicine, is by Flemish artist Peter Paul Rubens (1577–1640).

Several great traditions of medicine have arisen from cultures east and west. Western medicine, being firmly rooted in science, has proven the most successful in its ability to increase life expectancy and promote healthy living. Its traditional founder is Hippocrates (c.460–c.370 BCE), the Greek physician in whose name young doctors still swear to do their best by their patients. Hippocrates's medicine was based on the four elements: earth, air, water and fire. These were all believed to exist within the body in the form of the four humours, and disease was a consequence of imbalances between them. Hippocrates pioneered the techniques of diagnosis by looking for clues or symptoms of the problem and then giving the appropriate treatment in accordance with a prognosis (a prediction of how the illness would progress). Hippocrates chose interventions that treated the symptoms, in the hope that these would allow the body to prevail at the critical stages of the sickness. Additionally, Hippocrates was an intuitive advocate of the curative powers of hygiene and a good bedside manner.

**KEY DEVELOPMENTS**

**The four humours were: blood, which carried air; watery phlegm; black bile, which contains cold earth; and yellow bile filled with fire. Each humour affected mood. A rush of blood made you sanguine with an airy outlook. Phlegmatic people were calm like deep water. Depression – or melancholy – came from too much black bile, while yellow bile made people bilious and hot-tempered.**

DISCOVERING METABOLISM **p.54** CIRCULATION OF BLOOD **p.56** GERM THEORY **p.104** ANTIBIOTICS **p.130**
THERMOMETERS **p.187** BIG DATA **p.214**

# Alchemy

C.200 BCE – C.1750 CE

**KEY SCIENTISTS:** JABIR IBN HAYYAN • PARACELSUS • HENNIG BRAND ALBERTUS MAGNUS

Alongside the cerebral ponderings of the Greek natural philosophers, the alchemists offered a more hands-on approach to investigating nature. The best explanation for the root of 'alchemy' is 'from the land of Al-Khmi', an Arabized version of an ancient name for Egypt. *Khmi* means 'land of black earth', which relates to the fertile lands of the Nile Delta, where alchemy set down roots in the second century BCE. To the modern observer, alchemists were more like wizards than scientists. They made no distinction between magic and science, and invoking spirits with muttered chants during experiments was a crucial part of the process. Additionally, their goal was riches and power, and so alchemists were deliberately secretive. The work of Jabir ibn Hayyan (721–813 CE) was so hard to understand that his name gave rise to the word 'gibberish'. Nevertheless, alchemists developed much of the apparatus and glassware needed for precision research into chemicals.

**KEY DEVELOPMENTS**

**Learning about nature was a by-product of the alchemists' main goals, which were to get rich and live forever. They sought the Philosopher's Stone, a magical substance that transformed cheap materials into gold, and the Elixir, a spirit that cured all ills and bestowed immortality. Gunpowder was discovered by accident as Chinese alchemists sought everlasting life, while a greater understanding of compounds, elements and reactions arose from the search for the Stone.**

*The Alchemist* (after 1558), engraving by Pieter Bruegel the Elder (c.1525–1569).

→ GAS LAWS **p.60** SCIENTIFIC PROCESS **p.182** STANDARD MEASUREMENTS **p.185**

# Islamic Science

**KEY SCIENTISTS:** AL-RAZI • AL-HAYTHAM • AL-BIRUNI

Persian scientist-philosopher Abu Bakr Muhammad ibn Zakariyya al-Razi – commonly known as al-Razi – was widely regarded as the greatest physician of the Islamic world.

**KEY DEVELOPMENTS**

**The work of Islamic researchers – brought to Western Europe by returning crusaders – had a profound influence on people like Albertus Magnus (1199–1280) and Roger Bacon (c.1220–c.1292), both monks-cum-scientists. In their time, Europe was in the grip of the Scholastic tradition that sought to explain phenomena using the received wisdom of Aristotle, which had been incorporated into Church doctrine. Anyone who strayed too far from that doctrine risked being branded a sorcerer.**

The teachings of Islam attached great importance to education, and that is one reason why the focus of world science was shifting towards the Middle East by the eighth century CE. Baghdad's House of Wisdom took over from Alexandria's Great Library as the world centre of learning. Islamic alchemists expanded the subject of alchemy, the focus of which moved from the search for wealth and eternal life (although not completely) towards answers to practical commercial problems. These included how to preserve perfumes and create long-lasting pigments and glazes for ceramics – most notably blues, which were much in demand across the ancient world.

These goals required precision measurements and clear records and took big ste
towards modern chemistry. One notabl
alchemist of this time was al-Razi (854
925), who lived in the hills above Teh
He is credited with coining the term
*al-kuhl*, meaning the 'essence' or 's
a substance, from which came the
word 'alcohol'. Similarly, the En
'alkali' – any substance that ne
an acid – derives from the Ara
meaning a mixture of lime a

CAMERA OBSCURA **p.46** CIRCULATION OF BLOOD **p.56**

# Alchemy

C.200 BCE – C.1750 CE

**KEY SCIENTISTS:** JABIR IBN HAYYAN • PARACELSUS • HENNIG BRAND ALBERTUS MAGNUS

Alongside the cerebral ponderings of the Greek natural philosophers, the alchemists offered a more hands-on approach to investigating nature. The best explanation for the root of 'alchemy' is 'from the land of Al-Khmi', an Arabized version of an ancient name for Egypt. *Khmi* means 'land of black earth', which relates to the fertile lands of the Nile Delta, where alchemy set down roots in the second century BCE.
To the modern observer, alchemists were more like wizards than scientists. They made no distinction between magic and science, and invoking spirits with muttered chants during experiments was a crucial part of the process. Additionally, their goal was riches and power, and so alchemists were deliberately secretive. The work of Jabir ibn Hayyan (721–813 CE) was so hard to understand that his name gave rise to the word 'gibberish'. Nevertheless, alchemists developed much of the apparatus and glassware needed for precision research into chemicals.

**KEY DEVELOPMENTS**

**Learning about nature was a by-product of the alchemists' main goals, which were to get rich and live forever. They sought the Philosopher's Stone, a magical substance that transformed cheap materials into gold, and the Elixir, a spirit that cured all ills and bestowed immortality. Gunpowder was discovered by accident as Chinese alchemists sought everlasting life, while a greater understanding of compounds, elements and reactions arose from the search for the Stone.**

*The Alchemist* (after 1558), engraving by Pieter Bruegel the Elder (c.1525–1569).

GAS LAWS **p.60** SCIENTIFIC PROCESS **p.182** STANDARD MEASUREMENTS **p.185**

# Islamic Science

**KEY SCIENTISTS:** AL-RAZI • AL-HAYTHAM • AL-BIRUNI

Persian scientist-philosopher Abu Bakr Muhammad ibn Zakariyya al-Razi – commonly known as al-Razi – was widely regarded as the greatest physician of the Islamic world.

**KEY DEVELOPMENTS**

**The work of Islamic researchers – brought to Western Europe by returning crusaders – had a profound influence on people like Albertus Magnus (1199–1280) and Roger Bacon (c.1220–c.1292), both monks-cum-scientists. In their time, Europe was in the grip of the Scholastic tradition that sought to explain phenomena using the received wisdom of Aristotle, which had been incorporated into Church doctrine. Anyone who strayed too far from that doctrine risked being branded a sorcerer.**

The teachings of Islam attached great importance to education, and that is one reason why the focus of world science was shifting towards the Middle East by the eighth century CE. Baghdad's House of Wisdom took over from Alexandria's Great Library as the world centre of learning. Islamic alchemists expanded the subject of alchemy, the focus of which moved from the search for wealth and eternal life (although not completely) towards answers to practical commercial problems. These included how to preserve perfumes and create long-lasting pigments and glazes for ceramics – most notably blues, which were much in demand across the ancient world.

These goals required precision measurements and clear records and took big steps towards modern chemistry. One notable alchemist of this time was al-Razi (854–925), who lived in the hills above Tehran. He is credited with coining the term *al-kuhl*, meaning the 'essence' or 'spirit' of a substance, from which came the English word 'alcohol'. Similarly, the English word 'alkali' – any substance that neutralizes an acid – derives from the Arabic *al qaliy*, meaning a mixture of lime and water.

CAMERA OBSCURA **p.46** CIRCULATION OF BLOOD **p.56** SCIENTIFIC PROCESS **p.182** 

# The Renaissance

c.1400 – c.1550

**KEY SCIENTISTS:** LEONARDO DA VINCI • MICHELANGELO
NICOLAUS COPERNICUS

The Renaissance was a flowering of science, art and culture that began in the mercantile cities of Italy in the fifteenth century and spread across Europe. The term literally means 'rebirth', as if somehow Europe was regaining something lost. It began a new flow of knowledge from the Classical Greek world to the Islamic Middle East and then back along trade routes to Western Europe. The early Renaissance was dominated by polymaths – experts in many fields who were inspired to test old ideas, bend the rules and try new things. They included Leonardo da Vinci (1452–1519) who, while remembered mostly as a painter, was also a prolific inventor who sketched tanks and flying machines. Michelangelo (1475–1564) used art to break taboos while hiding in plain sight. For example, *The Creation of Adam* in the Sistine Chapel is an anatomical diagram of the human brain, with God and his angels picking out elements of the internal structure. Church doctrine forbade the dissection of dead bodies, and so this painting within the Vatican was a highly subversive image.

**KEY DEVELOPMENTS**

Even at the height of the Renaissance, Church teaching still placed Earth at the centre of the Universe, just as Aristotle had taught. In 1543, Polish astronomer Nicolaus Copernicus (1473–1543), who also worked for the Church, dared to suggest the unthinkable: Earth was just another planet that moved around the Sun. He also had the observational data to prove it. Keenly aware of the dangers of promoting such heretical ideas, Copernicus held back on publishing them until he had only a few days left to live.

Michelangelo's *The Creation of Adam* (c.1512), part of the fresco on the ceiling of the Sistine Chapel in the Vatican.

REFRACTION AND RAINBOWS **p.48** FINDING HOMOLOGUES **p.50** SCIENTIFIC PROCESS **p.182**

# The Scientific Revolution

**KEY SCIENTISTS:** WILLIAM GILBERT • ISAAC NEWTON • GALILEO GALILEI
ROBERT BOYLE • JOHANNES KEPLER

The seventeenth century brought a new rigour to scientific investigation, and this marks the start of modern science. Step by step, researchers around the world began to build new work on the discoveries of others, resulting in significant shifts in human understanding of the world. For example, William Gilbert (1544–1603) revealed the power of experiment, placing a compass on a magnetic sphere to show that it behaved in the same way as it did on the surface of Earth. His conclusion was that Earth is a magnetic sphere, too. Galileo Galilei (1564–1642) began using measurements and mathematics to uncover universal laws of motion. Johannes Kepler (1571–1630) did the same with his laws of planetary motion, which helped to cement in place the fact that Earth orbited the Sun, and the Moon moved around Earth. It was this system that led Isaac Newton (1642–1727) to formulate his laws of motion and gravity in the 1660s. Similarly, in the field of biology and medicine, William Harvey (1578–1657) used a scientific approach to prove that blood circulates within the body.

**KEY DEVELOPMENTS**

**Robert Boyle (1627–1691), an Anglo-Irish scientist, was responsible for the foundation of modern chemistry with his 1661 book *The Sceptical Chymist*. In it, Boyle refuted the magic and mysticism that dominated alchemy, and advocated building a view of substances according to systematic observations. Boyle's work paved the way for atomic theory and the laws of thermodynamics.**

Johannes Kepler, the celebrated German astronomer who banished medieval misconceptions about the movements of the planets.

PENDULUM LAW **p.52** ACCELERATION UNDER GRAVITY **p.55** CIRCULATION OF BLOOD **p.56** GAS LAWS **p.60**
SCIENTIFIC PROCESS **p.182** GRAPHS AND COORDINATES **p.183**

# The Rise of the Scientific Institution

**KEY SCIENTISTS:** BLAISE PASCAL • MARIN MERSENNE • RENÉ DESCARTES ROBERT HOOKE • EDMOND HALLEY • BENJAMIN FRANKLIN

A nineteenth-century engraver's impression of a meeting of the Lunar Society at Heathfield, Birmingham, England.

Communication is a crucial part of the scientific method, with researchers sharing results so that others may critique them, replicate experiments and investigate further. Today, scientists in all fields are networked together with counterparts across the globe via learned academies, colleges and societies for this very purpose. All these institutions can be traced back to two prototypes in Paris and London in the mid-seventeenth century. The French one was the Académie Parisienne, an informal group led by Marin Mersenne (1588–1648), a friar who disseminated the latest research chiefly by letter among a select group of scholars. Mersenne's group became the Académie des sciences in 1666. In England, Robert Hooke (1635–1703), Edmond Halley (1656–1742) and Robert Boyle had a similar collaboration called the Invisible College, and went on to be founding figures of the Royal Society, the world's oldest academy of science, founded in 1660.

**KEY DEVELOPMENTS**

**Alongside the national science academies, less grandiose organizations emerged in the eighteenth century that fed public interest in the latest exciting scientific developments. For example, engineers and industrialists such as James Watt (1736–1819), the steam engine pioneer, mingled with scientists such as Benjamin Franklin (1706–1790) in the Lunar Society of Birmingham, so called because members met each full moon.**

HOOKE'S LAW **p.62** THE SPECTRUM **p.66** THE FLYING BOY **p.68** WEIGHT OF THE EARTH **p.74** UNIVERSAL GRAVITATION **p.158** SCIENTIFIC PROCESS **p.182**

# The Birth of Chemistry

**KEY SCIENTISTS:** JOSEPH BLACK • JOSEPH PRIESTLEY • ANTOINE LAVOISIER

The early eighteenth century saw a boom in research into minerals, mostly in medical schools where students searched for new therapeutic substances. In the 1750s, Scottish medical student Joseph Black (1728–1799) discovered a gas while searching for a cure for kidney stones. He investigated further and named his discovery 'fixed air' because it seemed to be locked inside a mineral 'earth' and was released by heat. Today we know fixed air as carbon dioxide. Black thus became the world's first experimental chemist. In 1766, Henry Cavendish (1731–1810) found 'flammable air' (hydrogen) by reacting metals with acids. Daniel Rutherford (1749–1819) isolated a stifling, unreactive portion from the air, which he called 'phlogisticated air' (meaning it was unable to burn; today's name for it is nitrogen). The study of gases, or pneumatic chemistry, reached its peak in 1774 when Joseph Priestley (1733–1804) announced 'dephlogisticated air', which promoted burning and is now known as oxygen. These discoveries allowed later chemists to show that the ancient theory that there were only four elements – water, earth, air and fire – was a misconception.

**KEY DEVELOPMENTS**

**In the early days of chemistry there was no understanding that chalk, limestone and seashells, for example, were all basically the same material (calcium carbonate), and so the list of materials to be studied was almost inexhaustible. The discovery of distinct types of air, or gas, led to the further discovery that all materials, both natural and artificial, are made from a unique combination of simple substances now known as the elements.**

This famous James Gillray (1756–1815) comic etching is entitled 'Scientific Researches! New Discoveries in Pneumaticks!' It shows the increasing public interest in chemistry at the end of the eighteenth century, as Humphry Davy (right) and Thomas Garnett administer some of the new 'airs' to a volunteer.

GAS LAWS **p.60** OXYGEN **p.72** CONSERVATION OF MASS **p.76** ELECTROLYSIS **p.84** VITALISM **p.92**
ATOMIC NUCLEUS **p.126** ATOMIC THEORY **p.159** THE PERIODIC TABLE **p.162** DISTILLATION **p.204**

# Science and the Industrial Revolution

**KEY SCIENTISTS:** ALESSANDRO VOLTA • HUMPHRY DAVY • SADI CARNOT
HANS CHRISTIAN ØRSTED • JAMES PRESCOTT JOULE • LORD KELVIN

During the Industrial Revolution modern economies emerged, first in the United Kingdom and then in northern Europe and North America. The changes resulted in the biggest social upheaval since the development of agriculture.

The revolution was itself the result of a scientific upheaval that revealed a deeper understanding of the nature of motion, force, heat and electricity. Driven to exploit newly forged maritime trading routes, engineers made use of each new scientific breakthrough: 'hot gas' (steam) was used to turn heat into motion, while improvements in metal-refining allowed for the mass production of machines that automated the work of spinners, weavers, millers and threshers. New professions emerged: miners, steel workers, potters, bargemen and rail workers. Newly discovered metals, such as molybdenum, aluminium and chromium, were used to develop alloys that resisted corrosion and were tough but lightweight, and their colourful compounds opened up a new industry in pigments and paints.

**KEY DEVELOPMENTS**

**The Industrial Revolution was powered by the steam engine, which used the heat of burning coal to create high-pressure steam that could push a piston back and forth in a cylinder. The first practical power system – the 'atmospheric engine' of 1712, by Thomas Newcomen (1664–1729) – was vastly improved by James Watt, whose engines were bigger and stronger. The internal combustion engine, which used the hot gas of an exploding fuel to do the work, was invented in 1807.**

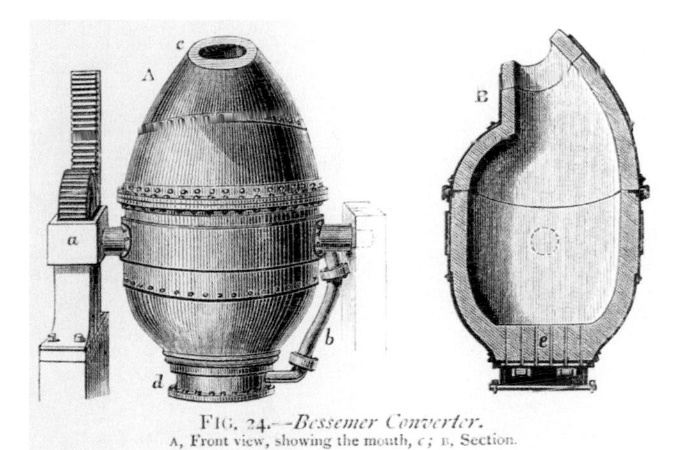


In the 1850s, English engineer Henry Bessemer (1813–1898) developed a system for making high-quality steel. Molten pig iron was poured into the pear-shaped vessels (A) and air was bubbled through it to burn away impurities.

PENDULUM LAW **p.52** WEIGHING THE AIR **p.58** LAWS OF THERMODYNAMICS **p.160** THERMOMETERS **p.187**

# Natural History and Biology

**KEY SCIENTISTS:** CARL LINNAEUS • CHARLES DARWIN • NICOLAS STENO GILBERT WHITE • ALEXANDER VON HUMBOLDT

Frontispiece from *The Natural History of Selborne* (1789) by Gilbert White (1720–1793), the pioneering English naturalist, ecologist and ornithologist.

Biology as a scientific field distinct from medicine was the last of the main disciplines to emerge from the Scientific Revolution. Of course, natural philosophers had addressed the subject prior to this, but the binomial classification system created by Carl Linnaeus (1707–1778) gave precision when describing species, and so made it possible for researchers to tackle the great diversity of life. This became a burgeoning activity in the early nineteenth century, when explorers like Alexander von Humboldt (1769–1859) and Charles Darwin (1809–1882) toured the world and returned with questions about why life varied so much from place to place but also displayed many of the same features and behaviours. Darwin famously used this inquiry to develop his theory of evolution by natural selection. One of Humboldt's legacies is the science of ecology, which investigates how wildlife communities interact with their environment. By the twentieth century, technological advances led to a new synthesis of biology that sees life as a phenomenon arising from chemistry and geology.

**KEY DEVELOPMENTS**

Niels Steensen (1638–1686), a Danish bishop, is better known by his Latinized name Nicolas Steno. Steno's work lies at the intersection between evolutionary biology, palaeontology and geology. In the 1660s he examined what he would have known as tongue stones due to their shape, but he recognized that they had the same serrated structures as the teeth of a shark. He concluded that these were in fact the petrified remains of a once-living creature – what is now understood to be a fossil.

DISCOVERY OF MICROORGANISMS **p.64** GERM THEORY **p.104** EVOLUTION BY NATURAL SELECTION **p.161**
CENTRAL DOGMA OF BIOLOGY **p.172** ENDOSYMBIOSIS **p.173**

# Geology and the Earth Sciences

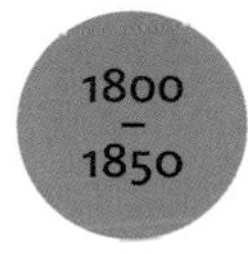


**KEY SCIENTISTS:** JAMES HUTTON • LOUIS AGASSIZ • CHARLES LYELL

The father of geology (the study of the physical Earth) is James Hutton (1726–1797), a Scottish farmer who observed local rock formations. Some of these formations were underground, others had been exposed to the surface by various engineering projects. Hutton was not the first to notice that the rocks were present in layers, or strata, but it was he who proposed the theory of uniformitarianism, according to which rocks are made at Earth's surface and deeper rocks are older than those layered on top. The older rocks were made by the same kinds of processes as those we see changing rocks at the surface today. Therefore, we can surmise what the surface was like in the distant past by examining the composition of ancient rocks formed at different times. Hutton's work led to the conclusion that Earth is very old indeed, an idea that had a profound influence on Charles Darwin among others. In addition, in 1840 Louis Agassiz (1807–1873) used Hutton's theory to show that much of Earth had once been covered by glaciers, the first evidence of ice ages.

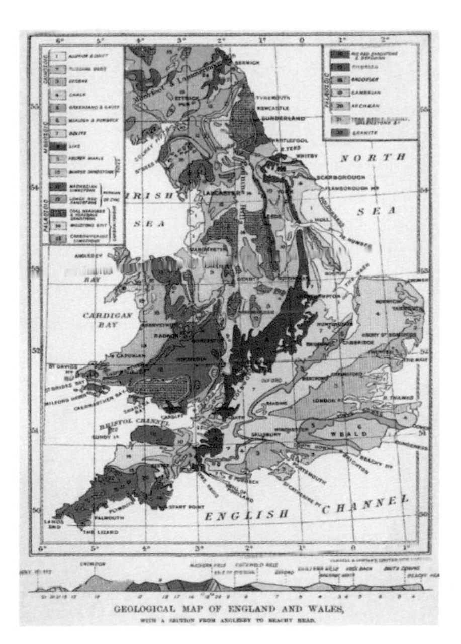


This nineteenth-century geological map shows some of the different rock strata that comprise the land mass of England and Wales.

**KEY DEVELOPMENTS**

**The geologists who followed Hutton pieced together a complete picture of how rocks formed, known as the rock cycle. Molten material deep in Earth cools into solid rock as it is pushed to the surface. Extreme pressures inside mountains may alter the rocks' make-up and then erosion eventually breaks apart all rocks into grains. These grains form sediments that petrify into new types of rock. Finally, the rocks are pulled down into the deep, molten Earth, ready to recommence the cycle.**

PROVING EXTINCTION **p.82** EVOLUTION BY NATURAL SELECTION **p.161** PLATE TECTONICS **p.164**
SEISMOMETERS **p.196** RADIOCARBON DATING **p.197**

# Electricity

1800 – 1830

**KEY SCIENTISTS:** ALESSANDRO VOLTA • BENJAMIN FRANKLIN MICHAEL FARADAY

Electricity has been studied since ancient times. In the seventh century BCE, Thales examined the way amber could attract and repel fluff, and hence the phenomenon was named after the Greek word *elektra*, meaning 'amber'. By the mid-eighteenth century CE, such charges of static electricity were being stored in foil-coated glass containers known as Leyden jars. It was one of these devices that Benjamin Franklin considered filling with charge from a lightning bolt. He never did (although others died trying), but he propagated the idea that electrical charges could be positive and negative.

In 1800, Alessandro Volta (1745–1827) created the first modern battery, which used a chemical reaction to create a continuous flow of electric charge – otherwise known as a current. The battery, of which there are various designs, allowed researchers to create a continual transfer of energy to make light and to drive machines.

**KEY DEVELOPMENTS**

**In 1820, the link between electricity and magnetism was revealed, and one year later Michael Faraday (1791–1867) harnessed the two forces to create a motor. The 1821 model was a very primitive prototype that used the forces of repulsion and attraction between a wire and a magnet to create a continuous circular motion. Ten years later, Faraday found that moving a wire through a magnetic field induced a current in the wire. So as well as inventing the motor, Faraday also invented the electricity generator.**

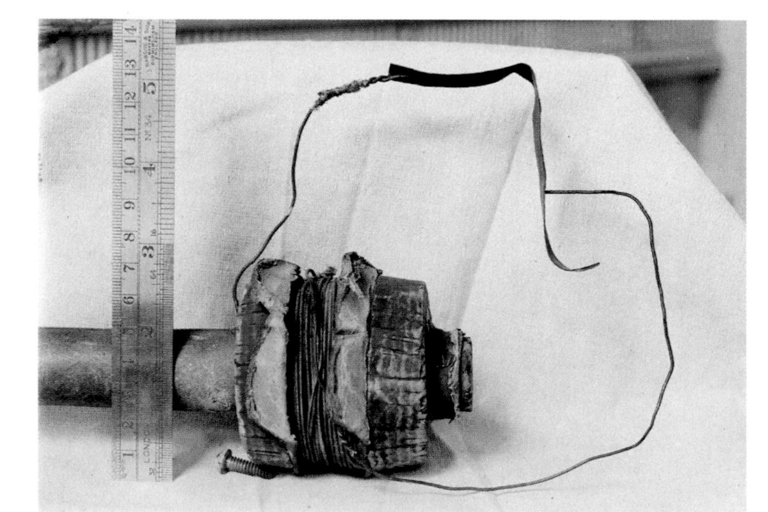

The coil of wire through which Michael Faraday created what he described as 'a very distinct though small' spark.

THE FLYING BOY **p.68** ANIMAL ELECTRICITY **p.78** ELECTROLYSIS **p.84** ELECTROMAGNETIC UNIFICATION **p.88**

# Cell Theory

**KEY SCIENTISTS:** ROBERT HOOKE • MATTHIAS SCHLEIDEN
THEODOR SCHWANN

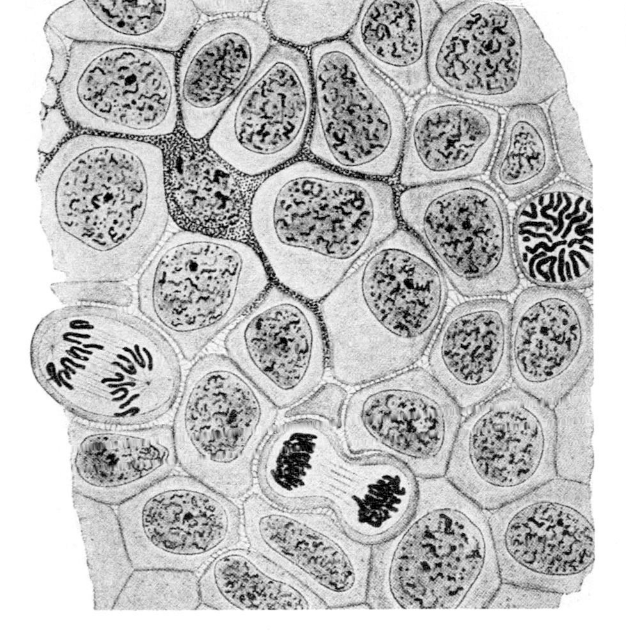

Highly magnified tissue cells from the skin of a salamander, taken from *Biology and its Makers* by William A. Locy, 1908.

**KEY DEVELOPMENTS**
As microscope technology improved, it provided an ever-growing supply of evidence for cell theory.
In the 1850s, researchers saw how cells divided in two by a complex set of steps known as mitosis. This led to interest in how information was transmitted from the original cell to new ones, and whether that was linked with inheritance from parent to offspring.
The answers led to the whole new field of genetics and the theory of evolution by natural selection.

In 1665, Robert Hooke undertook some of the first microscopic observations of living things and recorded his findings in a book, *Micrographia*. Among his discoveries were small structural units in cork. Hooke likened them to the living quarters of monks – small enclosures known as 'cells' (prisoners of the period were kept in dungeons). The term stuck, and about 170 years later cells were shown to be the basis of all life. This is cell theory, which is largely attributed to the work of two Germans: Matthias Schleiden (1804–1881), a plant expert, and Theodor Schwann (1810–1882), an animal physiologist. The theory has three tenets. First, all organisms are composed of one or more cells. (That means that viruses are not organisms and therefore not alive.) Second, the cell is the structural unit upon which the body is based. Third, and most profoundly: every cell arises from a pre-existing cell. Cell theory sounded the death knell for the earlier theory of spontaneous generation, which proposed that life, especially in its smallest forms, somehow emerged from the non-living ooze of decay.

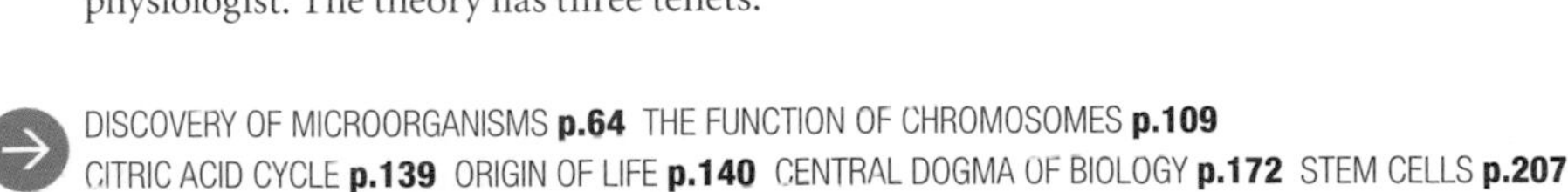
DISCOVERY OF MICROORGANISMS **p.64** THE FUNCTION OF CHROMOSOMES **p.109**
CITRIC ACID CYCLE **p.139** ORIGIN OF LIFE **p.140** CENTRAL DOGMA OF BIOLOGY **p.172** STEM CELLS **p.207**

# Public Health

**KEY SCIENTISTS:** JOHN SNOW • IGNAZ SEMMELWEIS • JOHN GRAUNT

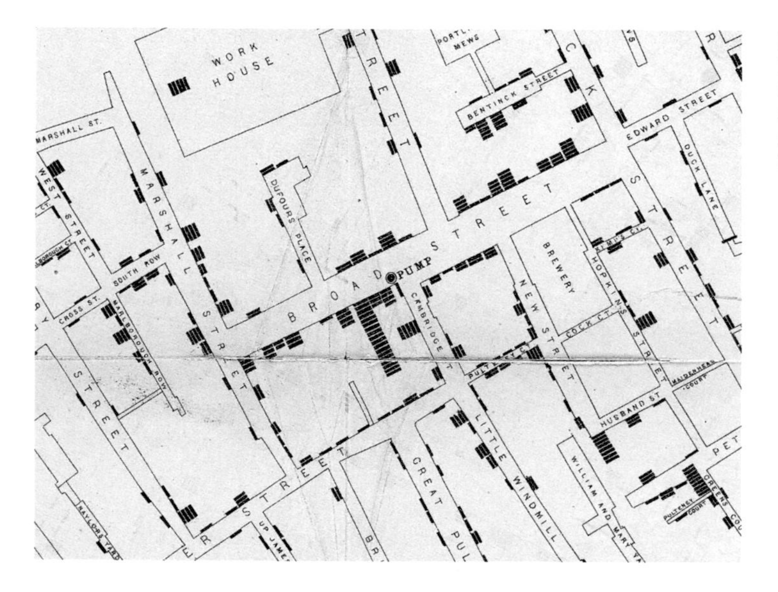


Map of Soho used by John Snow to identify the source of cholera in that part of central London.

The urbanization of the world crowded millions of people together in unsanitary environments and by the nineteenth century this had created the perfect conditions for diseases, most notably cholera, which is caused by bacteria in drinking water contaminated with sewage. However, in the 1850s this was unknown. Instead, the cause was thought to be either miasma – a fetid quality of 'bad air' – or a consequence of the abdomen getting cold. When cholera hit Soho, then a slum region of London, local doctor John Snow (1813–1858) began to investigate, mapping the victims and questioning them about their daily activities. He found that all local households with cholera sufferers collected water from the same public pump. He petitioned the authorities to seal the source, and when they did the Soho outbreak ended. John Snow is remembered as one of the founding figures of epidemiology, the science of how diseases impact a community.

**KEY DEVELOPMENTS**

John Snow's detective work made a strong link between hygiene and disease, which was confirmed in the following decades by the germ theory of disease. In 1844 Hungarian physician Ignaz Semmelweis (1818–1865) showed with clear statistical evidence that when medical staff washed their hands before attending births, new mothers were far less likely to die from infections such as so-called 'childbed fever'. Infuriatingly, Semmelweis's data were ignored until after his death.

DISCOVERY OF MICROORGANISMS **p.64** GERM THEORY **p.104** ANTIBIOTICS **p.130**
GRAPHS AND COORDINATES **p.183** CLINICAL TRIALS **p.208**

# The New Physics

**KEY SCIENTISTS:** MAX PLANCK • ALBERT EINSTEIN • NIELS BOHR

**KEY DEVELOPMENTS**

Einstein's most memorable contribution is the equation $E=mc^2$. This piece of mathematics relates how energy (E) is equivalent to mass (m). The scaling factor is the square of the speed of light (c). That is a very large number with 17 digits, so a small amount of mass contains a vast amount of energy, and that is what gives nuclear reactions their great power.

For his work on quantum theory, German scientist Max Planck (photographed here in 1933) was awarded the 1918 Nobel Prize in Physics.

In 1878, Max Planck (1858–1947) was advised by his university tutor to avoid specializing in physics, because 'almost everything has been discovered'. Planck did not listen, and by the turn of the century he made the founding discovery in quantum physics, namely that light and other radiation are emitted and absorbed by atoms in packets with discrete and unmodifiable quanta of energy. This evolved into a probabilistic understanding of atoms and subatomic particles, where behaviours and properties were not fixed in the same way as the old-school physicists were so certain about. These developments also threw into stark relief a contradiction between the laws of mechanics that govern motion and the laws of electromagnetism which rule over light and radiation. Albert Einstein (1879–1955) wondered why light seems to travel at the same speed even when the sources of it are travelling at different speeds. The result was the general theory of relativity, according to which space and time are not immutable and absolute but dynamic and capable of acceleration and deceleration, even though those effects may be imperceptible at a local level.

SPEED OF LIGHT **p.98** DISCOVERY OF THE ELECTRON **p.114** WAVE-PARTICLE DUALITY **p.128**
UNCERTAINTY PRINCIPLE **p.166** GEIGER-MÜLLER TUBE **p.191**

# The Size of the Universe

**KEY SCIENTISTS:** EDWIN HUBBLE • HENRIETTA SWAN LEAVITT
VESTO SLIPHER • J.J. THOMSON • MARIE CURIE • ERNEST RUTHERFORD

In the first 30 years of the twentieth century, physics expanded our view of the scale of the Universe, both in terms of its immense size and in relation to the minute particles deep inside atoms. Particle physicists found that the subatomic structure of atoms – their electrons, protons and neutrons – could be used to explain chemical and physical properties observed on the macro scale. An electron's configurations were linked to an atom's reactivity, density and conductivity, for example, while the structure of the nucleus was linked to radioactivity, and gave the atom most of its mass. At the other end of the scale, astronomers became able to measure the distances to stars with some confidence and found that our galaxy is just a tiny part of the Universe separated from neighbouring galaxies by several light years. The Milky Way is part of a galaxy cluster, and in turn forms a supercluster with other clusters, which form filaments and sheets in space that surround vast empty voids, akin to the soapy surface of a bubble.

**KEY DEVELOPMENTS**

**The surprising fact about atoms is that they are all largely empty space. The simplest view of their structure is to imagine a dense nucleus at the centre with a shell of electrons moving around it. If the shell had the diameter of a soccer stadium, then the nucleus would be the size of a football in the centre of the pitch.**

Henrietta Swan Leavitt (1868–1921), the pioneering US astronomer whose work established and measured the link between the luminosity of stars and intergalactic distances.

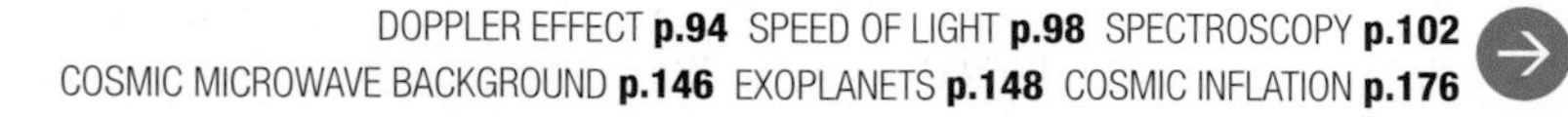
DOPPLER EFFECT **p.94** SPEED OF LIGHT **p.98** SPECTROSCOPY **p.102**
COSMIC MICROWAVE BACKGROUND **p.146** EXOPLANETS **p.148** COSMIC INFLATION **p.176**

# Science and the Public Good

**KEY SCIENTISTS:** J. ROBERT OPPENHEIMER • RICHARD FEYNMAN • ALAN TURING
WILLIAM SHOCKLEY

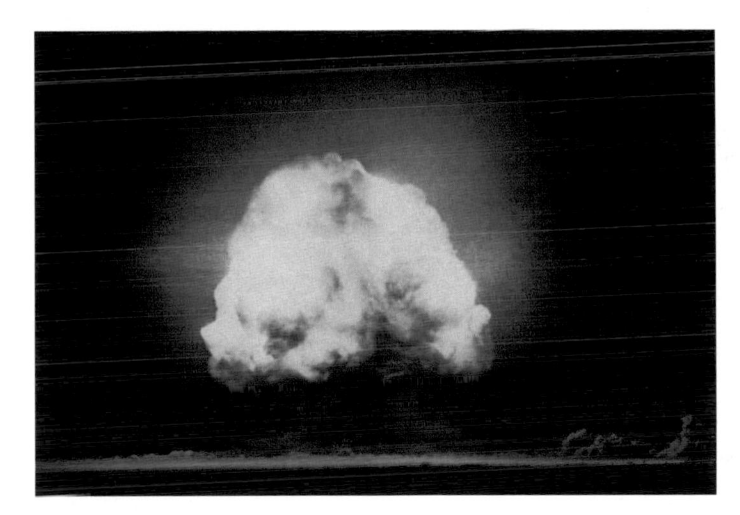

Code-named Trinity, the first detonation of a nuclear bomb was conducted by the US Army in the Jornada del Muerto desert, New Mexico, on 16 July 1945. Three weeks later, the US Air Force dropped atomic bombs on Hiroshima and Nagasaki, Japan, bringing an end to World War II.

Throughout World War II scientists strove to develop weapons with which to defeat their opponents. After the conflict, some of the technologies employed in the drive for victory were adapted for peaceful purposes. For example, radar technology, coupled with jet engines, both developed for war, have made air travel safe, reliable and within reach of ordinary people.

Additionally, research into radar wavelengths led to the invention of microwave ovens, and the pure silicon crystals used in receivers formed the base materials for semiconductors and microchips.

Computers were also invented in wartime, firstly as electromechanical calculators for crunching artillery targeting data in a hurry and also for cracking enemy codes. The first space vehicles were supersonic rockets designed as long-range missiles. The same technology has since been used to launch a host of satellites – some with robotic equipment, others with human crew – that have explored space and visited other worlds on our behalf.

**KEY DEVELOPMENTS**

**World War II was ended using nuclear weapons that unleashed immense explosions by splitting atoms in a lightning-fast chain reaction. To create such bombs, scientists needed to learn to build and control nuclear reactors. Today, reactors descended from this feat produce 10 per cent of the world's electricity. Nuclear reactors are also used to manufacture radioactive substances used in life-saving medical procedures.**

DISCOVERY OF ELECTROMAGNETIC WAVES **p.110** DISCOVERY OF RADIOACTIVITY **p.112**
NUCLEAR FISSION **p.136** TURING MACHINE **p.138**

# Electronics and Computation

**KEY SCIENTISTS:** ALAN TURING • JOHN VON NEUMANN • GEORGE BOOLE
WILLIAM SHOCKLEY • JACK KILBY

Technicians connecting the wiring of the ENIAC (Electronic Numerical Integrator and Computer), the first electronic general-purpose computer. It was financed by the US Army during World War II.

Electronics is a technology that controls the flow of electrons as current. In its most primitive form, electronics was based on thermionic valves, themselves an offshoot of light bulb technology. 'Solid-state' electronics was created in 1948 with the invention of a transistor switching device made from silicon. The silicon was doped with tiny amounts of other materials to make it a semiconductor that could flip between conducting a current and blocking its passage thousands of times a second. This switching is used as the physical manifestation of a digital code of 1s (for 'on') and 0s (for 'off'). In the 1950s, techniques for making solid-state devices cheaper and smaller saw the development of the integrated circuit, in which all the components and the connections between them were made on the same small section, or chip, of silicon. To this day, microchips form the central processing unit (CPU) of any computer, be it a desktop, a washing machine or a Mars rover. Inputs enter from outside the device and are processed according to the rules set out in the programming into outputs, creating powerful and versatile tools.

**KEY DEVELOPMENTS**

**Classical computers use Boolean logic, a form of mathematics developed in 1854 by George Boole (1815–1864). Every Boolean calculation gives the answer 1 (true) or 0 (false). Instead of familiar operations, such as addition and multiplication, Boolean calculations give strange-looking results where two inputs of 1 can give an output of 1 or 0 but never 2. Transistors are connected in patterns to create logic gates that produce results calculated using a particular Boolean operation.**

TURING MACHINE **p.138** COMPUTER GRAPHICS FOR MODELLING **p.211** CLIMATE SIMULATION **p.212**
MACHINE LEARNING **p.213** BIG DATA **p.214**

# Genetics

1952–

**KEY SCIENTISTS:** GREGOR MENDEL • CHARLES DARWIN • WILLIAM BATESON

The concept of inheritance has been recognized since the dawn of history, but its mechanism was largely unclear until the middle of the twentieth century. The theory of evolution relies on genetic information being passed from parent to offspring, and Darwin proposed a version of a very old theory known as pangenesis. This entailed every body part – heart, brain, etc. – sending its own particle of information to the sperm and egg. Gregor Mendel (1822–1884) had already shown that characteristics were inherited as discrete units that he termed 'factors'. These factors were renamed 'genes', a term inspired by the work of Darwin, and their study became known as genetics.

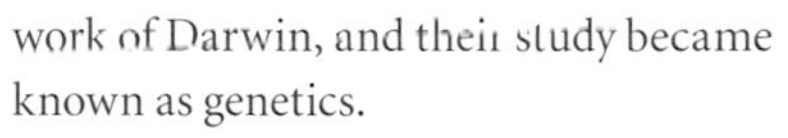

Modern genetics encompasses the study of deoxyribonucleic acid (DNA). In the early days, geneticists knew that DNA was present in the nucleus of a cell, along with the chromosomes, and in 1952 it was proven to be the agent of heredity. The following year, the structure of DNA was unlocked, and since then geneticists have been deciphering the code that it carries and relating that to the growth and function of the body.

The principles of heredity laid down by Gregor Mendel laid the foundations for the modern science of genetics.

**KEY DEVELOPMENTS**

One of the rules of genetic heredity is that information only passes from the gene to the organism, never in reverse. However, in the 1990s a related field appeared: epigenetics. Epigenetics examines the possibility that changes to the chemical wrappers that hold DNA – which vary according to conditions such as lifestyle, illness and famine – could be inheritable factors. There is evidence that hormones released from a significant life event experienced while pregnant can be transmitted to the foetus child and also on to any grandchildren.

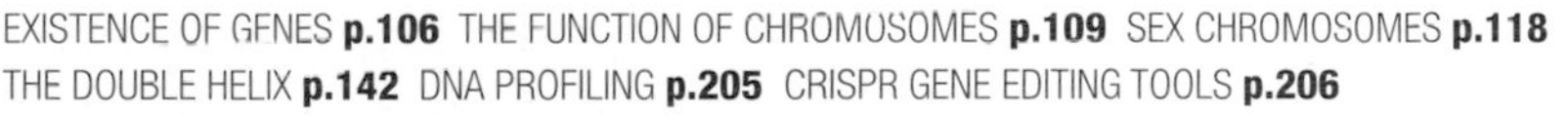

EXISTENCE OF GENES **p.106** THE FUNCTION OF CHROMOSOMES **p.109** SEX CHROMOSOMES **p.118**
THE DOUBLE HELIX **p.142** DNA PROFILING **p.205** CRISPR GENE EDITING TOOLS **p.206**

# The Space Race

**KEY SCIENTISTS:** KONSTANTIN TSIOLKOVSKY • ROBERT GODDARD
WERNHER VON BRAUN

American astronaut Edward H. White II, pilot of the Gemini IV four-day Earth-orbit mission, floats in zero gravity outside his spacecraft on 3 June 1965.

The first artificial object to enter space was a German V-2 rocket-powered bomb that made a suborbital test flight to an altitude of 176 kilometres (109 miles) in June 1944. V-2 technology, based largely on the earlier designs of US space pioneer Robert Goddard (1882–1945), was later used as the platform for Cold War superpowers to launch a Space Race. This competition was to become official in 1957, the International Geophysical Year, when both the United States and the Soviet Union aimed to launch satellites. The thinly veiled subtext was that the capability to place a spacecraft in orbit was also enough to hit any target on Earth with a rocket-powered missile. The Soviets won the first stage, launching Sputnik 1 in October 1957, and extended their lead as American efforts initially floundered before they scored astounding successes in the Apollo 'Moonshot'. Nevertheless, the first American spacecraft, Explorer 1, did succeed in making an actual scientific discovery in 1958. Its sensors confirmed that Earth's magnetic field spreading into space traps high-energy particles from the Sun in distinct belts that surround the planet.

**KEY DEVELOPMENTS**

**Most space launches do not carry living passengers, and after 60 years of exploration fewer than 600 people have visited space. The first life forms to enter space were fruit flies and moss that travelled in a V-2 commandeered by the Americans in 1947. The first large animal was a rhesus monkey, Albert II, in 1949; he was killed on landing. The first animals to return safely from orbit were the Russian dogs Belka and Strelka in 1960. In 1961, Yuri Gagarin (1934–1968) became the first human to fly into space.**

COSMIC RAYS **p.124** EXOPLANETS **p.148** ORIGIN OF THE SOLAR SYSTEM **p.179** TELESCOPES **p.189**
PLANETARY ROVER **p.215**

# Human Evolution

1960–

**KEY SCIENTISTS:** LOUIS LEAKEY • MARY LEAKEY • RICHARD LEAKEY

In 1972, a partial skeleton was unearthed in Ethiopia. The pelvic bones showed that the remains were of a bipedal hominid, a female who could walk upright on her legs, and who also had long arms that would have helped with climbing trees. The fossil was named Lucy, and it remains one of the earliest examples of a species that exhibits features close to those of our species and a significant divergence from the body plan of non-human apes.

The Rift Valley of East Africa has since provided evidence of several extinct human ancestors, building a picture of how our species, *Homo sapiens*, evolved. Lucy lived around 3.2 million years ago, and she made simple tools – the stone ones survive. A million years later, the region was home to *Homo habilis*, more human-like than Lucy but only just over 1 metre (3 feet) tall. This species is known for an extensive tool-kit.

What happened next is unclear. A later, larger species, *Homo erectus*, is known to have spread out of Africa around a million years ago, and was probably the ancestor of the Neanderthals. However, with new fossils recently being found across Africa, the precise lineage to modern humans, who spread to the rest of the world from around 70,000 years ago, has become somewhat obscure.

**KEY DEVELOPMENTS**
The conventionally agreed defining features of the human species include things like bipedality and large brains. It is likely that these features co-evolved. To stand on two legs requires the pelvis to rotate flat. That leads to a reduction in the size of the birth canal, which forces babies to be born small after a relatively short gestation. During the early stages of development, the brain is larger in proportion to the rest of the body, so ancestral human babies evolved big heads but were also entirely helpless and had to be carried by their parents – with arms no longer needed for walking.

British paleoanthropologist Mary Leakey (1913–1996) with the skull of a small primate, c.1940.

FINDING HOMOLOGUES **p.50** PROVING EXTINCTION **p.82** EXISTENCE OF GENES **p.106** ORIGIN OF LIFE **p.140** EVOLUTION BY NATURAL SELECTION **p.161** RADIOCARBON DATING **p.197** CLADISTICS AND TAXONOMY **p.209**

# Neuroscience and Psychology

1960–

**KEY SCIENTISTS:** SIGMUND FREUD • CAMILLO GOLGI • ERIC KANDEL DONALD HEBB

Psychology, the study of mental processes, has a nebulous foundation, but was firmly established by the mid-twentieth century and began to shed powerful light on human behaviour. Alongside it was neuroscience, the study of the form and function of nervous systems, especially at the cellular and biochemical level. However, these two subjects, so obviously related, were separated by a lack of evidence. Since there was no physical evidence linking mental activity with learning or behaviours, proponents of 'radical behaviourism' proposed that researchers should assume there was no link at all. Perhaps the mind was an artefact of something else, not a controlling principle ruling the body, they said. Despite this, the two fields developed apace, with the assumption that a link existed. The leading theory was that 'neurons that fired together wired together', which suggested that learning and memory were held in physical circuits of brain cells that were reinforced by being recalled. In the early 1970s, Eric Kandel (1929–) found the first physical link by showing how learning changed the chemistry of nerve cells in sea slugs.

**KEY DEVELOPMENTS**

Neuroscience and psychology still must tackle the same hard problems of consciousness: how, for example, can we study the internal sensations of the mind, such as the perception of colour or pain? These mental phenomena, known as qualia, are so far entirely private: even if everyone agrees that the sky is blue, if it were possible to perceive that colour as other people do, would all of them result in the same experience?

For his research on the physiological basis of memory storage in neurons, Austrian American Eric Kandel was one of the winners of the 2000 Nobel Prize in Physiology or Medicine.

ANIMAL ELECTRICITY **p.78** LEARNED RESPONSES **p.116** CLINICAL TRIALS **p.208**

# Environmental Sciences

**KEY SCIENTISTS:** JULIAN HUXLEY • RACHEL CARSON • PETER SCOTT DAVID ATTENBOROUGH

In 1962, Rachel Carson (1907–1964) wrote *Silent Spring*, a piece of long-form journalism that described how the natural world was being degraded by the activities of humans. Writing at the peak of the industrial age, Carson warned that the pesticide DDT and countless other chemicals carelessly left to pollute nature would lead to the unintended destruction of life. The title predicts that if this were to happen, life and its sounds would soon fade away. The book was instrumental in ensuing efforts to clean up the US chemical industry, and similar things happened elsewhere around the world. The book also helped to inspire the emergence of a grassroots environmental movement that valued the natural world and questioned whether industrial progress was worth the degradation of natural habitat. By 1970, the green movement had sufficient power for the US government to found the Environmental Protection Agency, which again was mirrored in other countries. Meanwhile, pressure groups such as Greenpeace and conservation non-governmental organizations, such as the World Wildlife Fund (WWF), became international players in protecting the environment.

**KEY DEVELOPMENTS**

Conservation is a complex field in which the needs of communities for food and space must be managed alongside protecting rare habitats and species in danger of extinction. Approaches vary accordingly, but the consensus on what if anything needs protecting is often based on information from the International Union for Conservation of Nature (IUCN), which maintains the 'Red List' of endangered animals. Each species on the list is classified according to the risk of its extinction, in the hope that measures will be taken to save it before it is too late.

British anthropologist Jane Goodall (1934–) has devoted her life to the study and conservation of chimpanzees in Tanzania.

ORIGIN OF LIFE **p.140** EVOLUTION BY NATURAL SELECTION **p.161** ANTHROPOGENIC CLIMATE CHANGE **p.178** CLIMATE SIMULATION **p.212**

# The Internet

c.1970–

**KEY SCIENTISTS:** PAUL BARAN • VINT CERF • BOB KAHN • TIM BERNERS-LEE

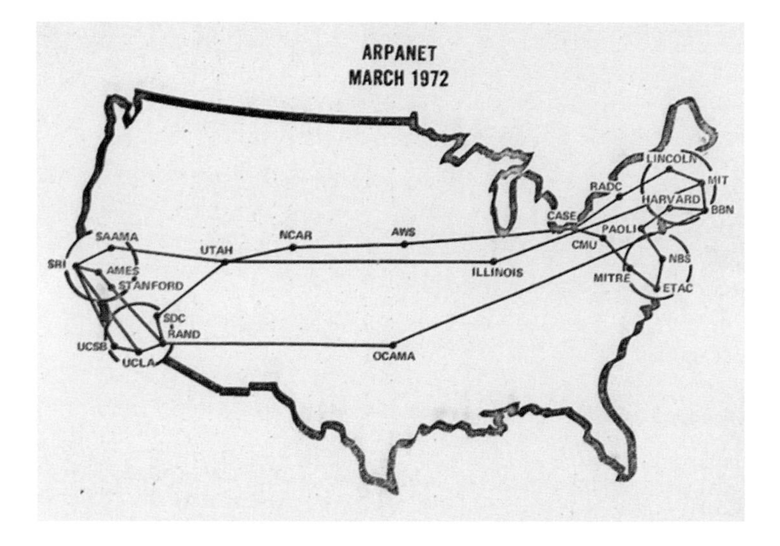


The extent of the Internet – then known as the Advanced Research Projects Agency Network (ARPANET) – after around 18 months in operation.

The Internet was initially a military project. In the nuclear age, commanders required seamless communications between their early warning systems and their high-tech arsenals, and had a growing dependency on computing to do it. All this innovation could be neutralized by simply cutting off commanders from their forces. So the US Department of Defense called for a more resilient system, and the result was packet switching. This is a system in which data are divided into packets, each of which travels through the communications network independently, potentially taking different routes. If packets fail to arrive, then the receiver's computer calls for them to be sent again, until the original message is reassembled. This technology was turned on in 1969 with connections between a few West Coast universities, but soon it opened up the possibility of a 'network of networks' or Internet, which required a computer user to have a single physical connection to the system to be able to communicate with any one of a growing number of users. Academics, especially scientists, were pioneers of the network, using the Internet to share large amounts of data with ease.

**KEY DEVELOPMENTS**

**The World Wide Web, created in the early 1990s by Tim Berners-Lee (1955–), was devised as a way to share information more effectively. The Web works by Internet-connected computers holding material that is viewable by anyone with a Web browser. Its incredible power is to allow people to find the information they want from sources they did not know existed – and do it within a fraction of a second.**

TURING MACHINE **p.138** COMPUTER GRAPHICS FOR MODELLING **p.211**

# The Universe is Missing

**KEY SCIENTISTS:** NICOLAUS COPERNICUS • ADAM RIESS • VERA RUBIN

American philosopher Thomas Kuhn (1922–1996) described how it was natural for science to enter periods of crisis, during which contradictions and mysteries abounded, before being reset by paradigm shifts. Examples of such shifts include the discovery of the heliocentric Solar System and the theory of relativity. As we entered the twenty-first century, the sciences – astronomy and cosmology especially – were in crisis. In the 1980s the field had to contend with the fact that the visible Universe made up only about one sixth of its matter. The rest was so-called dark matter that could only be detected by its gravity, but which is six times more abundant than normal matter. Next, at the close of the 1990s, astronomers found that matter – whether dark or not – made up only a quarter of the energy in the Universe. The rest was a dark energy somehow contained within the nothingness of empty space. So the centuries of scientific advance that explained the Universe with ever-increasing detail really referred to only 5 per cent of it; the other 95 per cent remains a mystery.

American astronomer Vera Rubin (1928–2016) while a student at Vassar College. In later life, her pioneering work on galaxy rotation rates helped to prove the existence of dark matter.

**KEY DEVELOPMENTS**

**In 2018 astrophysicist Jamie Farnes (1984–) proposed that dark energy and dark matter are parts of the same hypothetical material – 'dark fluid'. He suggested that, as the Universe expands, empty space is filled with a negative mass. (The gravity of objects with negative mass pushes them apart instead of pulling them together.) This negative mass would bind galaxies more strongly than normal mass, and that would match with observations that underlie the theory of dark matter. The idea is interesting but has few supporters.**

→ COSMIC MICROWAVE BACKGROUND **p.146** EXOPLANETS **p.148** DISCOVERY OF DARK ENERGY **p.150** DARK MATTER **p.175** COSMIC INFLATION **p.176** THE MANY WORLDS INTERPRETATION **p.177**

# Genetic Modification

1990–

**KEY SCIENTISTS:** RUDOLF JAENISCH • JENNIFER DOUDNA

By the twenty-first century, the technology behind genetic engineering was sufficiently advanced to make the leap from laboratory to store shelves, mostly in foods and medicines. That caught the attention of the public, and debate began over whether foods and the agriculture industry need to use genetically modified organisms (GMOs). One side argues that GMOs, most likely crop plants, are able to thrive in conditions previously too harsh for a rich harvest. They do that by drawing on capabilities, such as a tolerance to frosts or moulds, imparted by including a gene from an entirely different organism. They may also be designed to be resistant to a particular chemical that kills other organisms, and thus pesticide can be used more judiciously. The opposing arguments are that the genetic changes in GMOs are patented, and farmers are beholden to the patent owner in ways they aren't with natural plants. Additionally, plants hybridize more readily than animals, and environmentalists worry that GMOs could escape into the wild, using their genetic advantages to alter natural ecosystems. The debate continues.

**KEY DEVELOPMENTS**

**The genetic modification of microorganisms is less controversial. This is done mostly for medicinal purposes. For example, modified viruses are used to create vaccines, while bacteria and yeasts are engineered to produce useful biochemicals on industrial scales. One of the most successful modifications was made in 1978, when a new strain of E. coli was created, which produced the human form of insulin that is much more effective than previous forms at managing diabetes.**

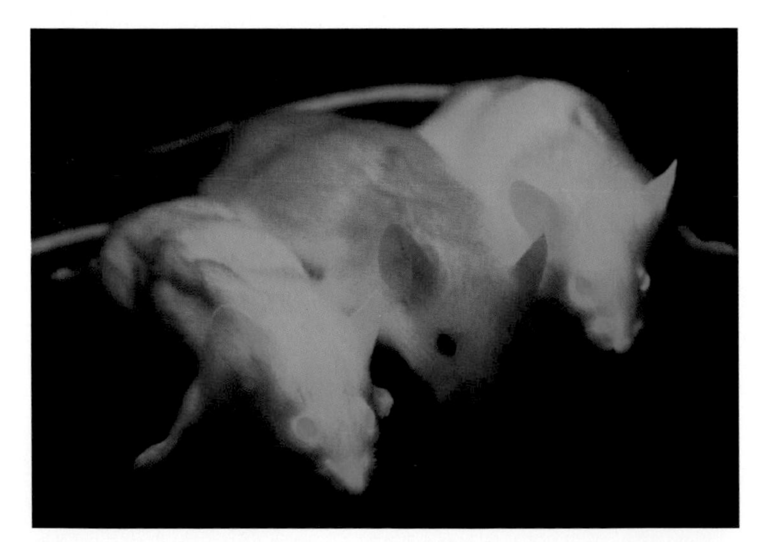

Genetic modification in mammals: two mice expressing enhanced green fluorescent protein under UV-illumination flanking one plain mouse from the non-transgenic parental line.

EXISTENCE OF GENES **p.106** THE DOUBLE HELIX **p.142** CENTRAL DOGMA OF BIOLOGY **p.172**
DNA PROFILING **p.205** CRISPR GENE EDITING TOOLS **p.206**

# String Theory

2000–

**KEY SCIENTISTS:** PETER HIGGS • EDWARD WITTEN

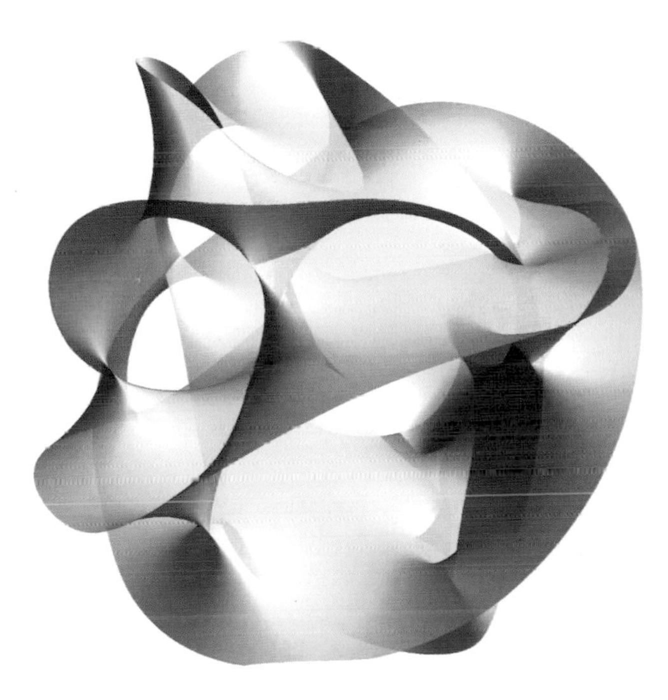

**KEY DEVELOPMENTS**
For every particle that we know about – electrons, photons, etc. – there might also be an as yet undiscovered (and very heavy) superparticle or sparticle – a superelectron or a super-photon. The Large Hadron Collider (LHC) – the particle accelerator near Geneva, Switzerland, used in 2012 by the European Organization for Nuclear Research (CERN) to find the Higgs boson – is being upgraded to harness the energy theorized to be needed to spot a sparticle.

This computer-generated image represents the surfaces of a shape that has four spatial dimensions – one more than the usual length, width and height. Some string theories deal with objects that have 11 dimensions.

The main problem with the two great theories of physics, quantum mechanics and relativity, is that neither can account for the other: for example, gravity cannot be explained in terms of quantum particles.

An attempt to reconcile the two, string theory proposes a deeper level of organization in which particles are strings that vibrate in numerous dimensions. Three of these dimensions are familiar to us – length, width and height – but, according to the theory, there are other dimensions that we cannot visualize.

The theory describes particles as a series of loops and one-dimensional strands. These objects manifest the multifarious properties of a particle as vibrations in the string that occur in eight compact spatial dimensions over and above the three that make up the everyday world.

String theory, or at least many of its versions, has a place for quantum gravity and so is a candidate for TOE, the Theory of Everything. However, so far no one has been able to go beyond the exquisite mathematics that underwrite string theory and find a way to test it in action.

DISCOVERY OF THE ELECTRON **p.114** ATOMIC NUCLEUS **p.126** THE STANDARD MODEL **p.174**
BUBBLE CHAMBERS **p.198** PARTICLE ACCELERATORS **p.199** NEUTRINO DETECTORS **p.201**

ANGELVM GALLIÆ CVSTODEM CHRISTVS PATRIÆ FATA DOCE
A CONVENTION NATIONALE
DU CAPITAINE GUY
POUR LA FRANCE
17
16
15
14
13
12

# EXPERIMENTS

# Buoyancy

C.250 BCE

**ARCHIMEDES:** *ON FLOATING BODIES* • SYRACUSE, SICILY, ITALY

GREEK PHILOSOPHERS **p.13**

**Key publications by Archimedes**
*Measurement of a Circle*
*Ostomachion*
*The Sand Reckoner*

Greek scientist Archimedes (c.287–c.212 BCE) is best known for having a big idea in the bath, leaping out of the water and shouting 'Eureka!' – meaning 'I have found it!' What Archimedes had found is now used to explain why some materials float and others sink.

Archimedes had been thinking about the purity of the gold in his king's new crown. As he relaxed into the brimming bath, he displaced some water. He realized that he could measure the volume of the crown by collecting the water it displaced in the same way. With an accurate volume, he could then compare the crown with a piece of gold of the same size. If the crown was pure, it would weigh the same as the metal ingot.

Archimedes compared the buoyancy of the items, hanging them on a balance beam and submerging them in water. The weight of an object pushes water away, but the water pushes back with a 'buoyant force'. If that force is greater than the weight, then the object floats; if the force is less than the weight, the object sinks. This is summed up by Archimedes' principle: a buoyant force exerted on an object is equal to the weight of the water displaced by that object. Archimedes knew that the buoyant forces acting on both items were equal, but the crown floated higher in the water than the gold, thus showing that the crown was not pure gold at all, but mixed with lighter – and cheaper – metals.

**ARCHIMEDES**
One of the most prolific mathematicians and inventors of the ancient world, Archimedes lived in Syracuse, a Greek city in Sicily. His achievements include calculating pi ($\pi$) to 3.1408, the most accurate figure yet. Legend links him with building weaponry to fend off Roman attacks, most notably heat rays that set warships alight using curved mirrors to focus beams of sunlight. Archimedes was killed when the Romans finally took the city in the Punic Wars.

This sixteenth-century German engraving illustrates the moment when, according to tradition, Archimedes established the principle that now bears his name.

STANDARD MEASUREMENTS p.185

# The Size of Earth

**ERATOSTHENES:** *DE MOTU CIRCULARI CORPORUM CAELESTIUM*
ALEXANDRIA, EGYPT

In the last few years of the third century BCE, Eratosthenes (c.276–c.194 BCE), a Greek polymath living in Alexandria, Egypt, came up with a way of measuring the circumference of Earth. The idea came to him when he heard of a very unusual well on Elephantine Island in the River Nile near the city of Syene (modern Aswan). It was said that at noon on midsummer's day the Sun cast no shadows in this deep well and was reflected in its waters. That told Eratosthenes that the Sun was directly overhead in Syene on that day, but he knew that at the same time in Alexandria it was not. So he erected a small pillar at Alexandria and measured the length of its shadow at noon. He then used that value along with the height of the pillar to construct a right-angle triangle, which allowed him to calculate that the Sun's rays hit Alexandria at an angle of 7 degrees. Eratosthenes saw that if his pillar reached all the way to the centre of Earth, then a similar line running into the planet from Syene would meet at this same angle, 7 degrees, and that meant that the distance between the two cities represented approximately one fiftieth of the distance around the world (one seventh of the 360 degrees in a circle). Eratosthenes consulted travellers to conclude that Syene was 5,000 Egyptian stadia away, and so Earth has a circumference of 250,000 stadia – 39,375 kilometres (24,466 miles). This was just 1.4 per cent under today's measurement: 40,076 kilometres (24,901 miles).

The well of Eratosthenes on Elephantine Island, Aswan, Egypt.

ANCIENT ASTRONOMERS **p.12** ISLAMIC SCIENCE **p.16**

**Key publications by Eratosthenes**
*Platonikos*
*Hermes*
*Erigone*
*Chronographies*
*Olympic Victors*

**ERATOSTHENES**

Eratosthenes was born in Cyrene, a Greek colony on the coast of what is now Libya. As a young man, he studied with leading philosophers of the time at the Platonic Academy in Athens. He made a name for himself as a poet and was asked to move to Alexandria and become one of the city's librarians. By the age of 35, he had been elevated to Chief Librarian and did much to make the Library of Alexandria the leading centre of learning in the ancient world.

SCIENTIFIC PROCESS **p.182** STANDARD MEASUREMENTS **p.185**

# Camera Obscura

C.1000 CE

**IBN AL-HAYTHAM:** *KITĀB AL-MANĀẒIR* • CAIRO, EGYPT

ISLAMIC SCIENCE **p.16**

**Key publications by al-Haytham**
*Treatise on Light*
*On the Configuration of the World*
*Doubts Concerning Ptolemy*
*Model of the Motions of Each of the Seven Planets*

Ibn al-Haytham (c.965–c.1040) – also known by the Latinized form Alhazen – was the founding figure of the field of optics, which is the study of the behaviour of light beams. Much of his work was based on – or more likely inside – the camera obscura. This term is Latin for 'dark chamber'. The camera obscura was a well-established phenomenon by the days of al-Haytham. The room – or tent – used for this purpose was completely dark, save for a single pinhole in one wall. Observing a solar eclipse from inside such a chamber, al-Haytham noted that the image of the Sun appeared upside down on the wall opposite the pinhole. He imagined how a set of beams from the outside scene converged on the pinhole and crossed over to create the inverted image visible inside. This led him to take a geometric approach to optics that always treated beams of light as straight lines which changed direction and angles as they were reflected off surfaces or refracted though them. Al-Haytham's view of light also demolished the ancient emission theory of vision, which proposed that the eyes sent out rays which bounced back to create an image. Instead, al-Haytham extended the intromission theory, according to which light given out from the Sun or another source reflects off an object and from there into the eye of the observer, making an image just as it did in the camera obscura.

**AL-HAYTHAM**

Born in Basrah (a city in modern Iraq) at the height of the Fatimid Dynasty, al-Haytham became a vizier or advisor to the local emir. He is said to have boasted that he had a plan to control the flood-waters of the River Nile, and was duly invited to Cairo by the caliph, al-Hakim. Pretty soon al-Haytham recognized the folly of his claims and, according to the more lurid biographies, feigned madness in order to avoid the caliph's wrath. Whatever the truth of the matter, al-Haytham was placed under house arrest for years, during which time he carried out his research into optics.

This engraving of 1752 illustrates the principle of light reflecting upside down on the wall of a camera obscura.

MICROSCOPES **p.188** TELESCOPES **p.189** PHOTOGRAPHY **p.192**

# Refraction and Rainbows

c.1300

**THEODORIC OF FREIBERG:** *DE IRIDE ET RADIALIBUS IMPRESSIONIBUS*
TOULOUSE, FRANCE

ISLAMIC SCIENCE **p.16**

In a rainbow, the incident white light is diffracted into its component colours to form arcs of (from outside to inside) red, orange, yellow, green, blue, indigo and violet.

**Key publications by Theodoric of Freiberg**
***De luce et ejus origine***
***De coloribus***
***De miscibilibus in mixto***
***De elementis corporum naturalium***

The link between the rainbow and sprays of water, not just rain, had been established by Seneca the Younger (c.4 BCE–65 CE). Noting that a rainbow always appears on the opposite side of the sky to the Sun, Seneca proposed that the colourful phenomenon was caused by light reflecting from each droplet individually. Al-Haytham had believed that rainbows came from the raindrops forming a concave mirror. However, Theodoric of Freiberg (c.1250–c.1310) concluded the debate by using a water-filled glass sphere to recreate the rainbow effect. As light entered the front surface of the sphere, it was refracted and dispersed into a narrow spectrum of colours which reflected off the rear inner surface. Leaving the front of the sphere, the spectrum refracted again. In total, light turns by 318 degrees, almost back towards the source of the light. Refraction is when a light beam deflects off its course as it transitions from one transparent medium (in this case, the air) into another (in this case, water). The angle of refraction depends on the relative speeds of light in each medium (it is slower in water, for example). Therefore every raindrop is bending the light in precisely the same way as all the others, and this cumulative action creates the rainbow across the sky.

**THEODORIC OF FREIBERG**

**Also known as Dietrich, Theodoric was the star pupil of Albertus Magnus, the influential German philosopher, scientist and theologian. Like his teacher, Theodoric became a Dominican friar, and had a long career in academia in France and Germany. As well as natural science, Theodoric was interested in metaphysics, the study of existence.**

MICROSCOPES **p.188** TELESCOPES **p.189** LASERS **p.195**

# Finding Homologues

C.1550

**PIERRE BELON:** *L'HISTOIRE DE LA NATURE DES OYSEAUX* • PARIS, FRANCE

One of the cornerstones of the theory of evolution is comparative anatomy, which is the study of two things: first, how a shared body plan of bones and limbs can evolve into different shapes to suit a variety of environments, for example how whales and bats have bodies built from skeletons made up of the same (homologous) bones. The second area of research is how different anatomies can take on similar forms that equip them for a similar way of life, such as dolphins, sharks and ichthyosaurs (extinct marine reptiles).

The founder of comparative anatomy was Pierre Belon (1517–1564), who published a series of books in the 1550s investigating the analogous anatomies of sea creatures. In 1555 he published *L'histoire de la nature des oyseaux*, which demonstrated in detail how the bird skeleton was homologous to (had much the same structure as) the human one.

Belon's work meant that the questions of why and how birds and mammals with the same basic skeletal platform then developed different anatomies were now open to interpretation. The discoveries of comparative anatomy gradually led to the view that life forms must develop, emerge or evolve from common ancestors.

**PIERRE BELON**

Born in Le Mans, France, Pierre Belon was apprenticed as an apothecary and went into service under the Bishop of Clermont. He became interested in zoology, the study of animals, and after entering university at Wittenberg he extended his research to plants. Belon had a short spell at a Paris medical school in the 1540s but never qualified as a doctor. Instead, he began extensive explorations of Europe and the Middle East, where he made his name as an anatomist. Back in Paris he was murdered by thieves.

THE RENAISSANCE **p.17** NATURAL HISTORY AND BIOLOGY **p.22** HUMAN EVOLUTION **p.33**

**Key publications by Pierre Belon**

*L'histoire naturelle des éstranges poissons marins, avec la vraie peincture & description du Daulphin, & de plusieurs autres de son espece* 1551

*De aquatilibus* 1553

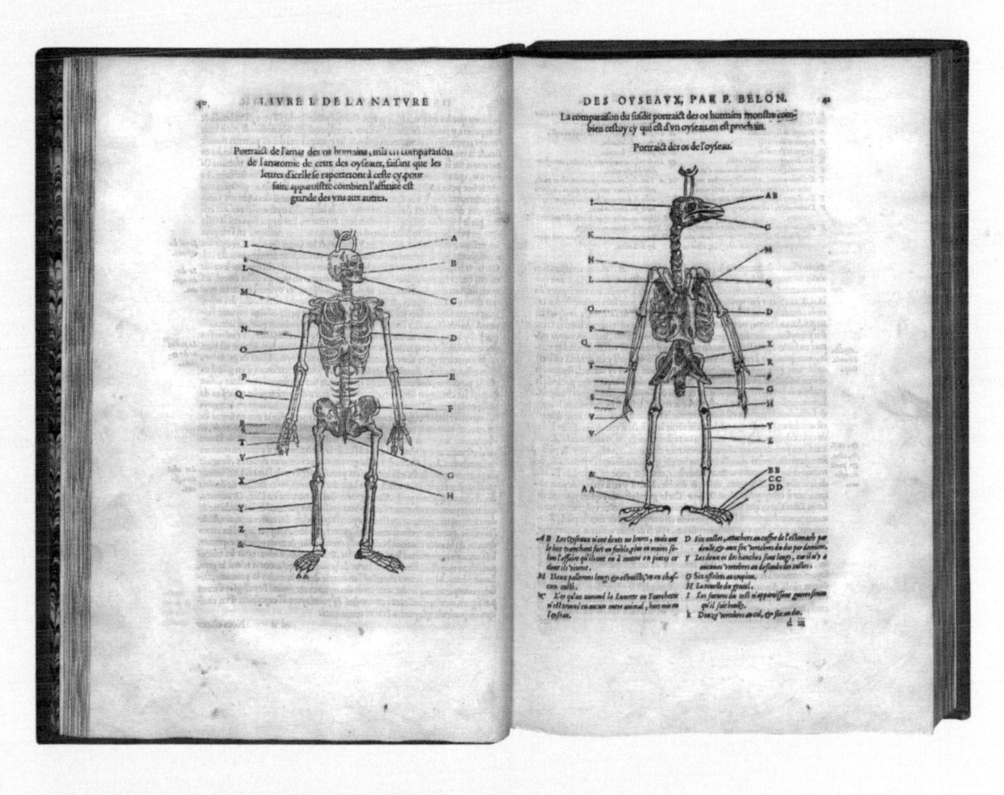
40. LIVRE I. DE LA NATVRE

Portraict de l'amas des os humains, mis en comparaison de l'anatomie de ceux des oyseaux, faisant que les lettres d'icelle se rapporteront à ceste cy, pour faire apparoistre combien l'affinité est grande des vns aux autres.

DES OYSEAVX, PAR P. BELON. 41

La comparaison du susdit portraict des os humains monstre combien cestuy cy qui est d'vn oyseau en est prochain.

Portraict des os de l'oyseau.

In highlighting the similarities between the skeletons of a human and a bird, this double-page spread from the first edition of Belon's chef d'oeuvre became a cornerstone of comparative anatomy.

EVOLUTION BY NATURAL SELECTION **p.161** ENDOSYMBIOSIS **p.173** CLADISTICS AND TAXONOMY **p.209**

# Pendulum Law

1583

**GALILEO GALILEI:** *DISCORSI E DIMOSTRAZIONI MATEMATICHE INTORNO A DUE NUOVE SCIENZE* • PISA, ITALY

**Key publications by Galileo**
*The Little Balance* 1586
*Mechanics* c.1600
*The Operations of Geometrical and Military Compass* 1606

A pendulum is a weight – a bob – that swings on a cord or rod. Pendulums had long been seen to be sensitive to motion and so were used in early technologies such as vibration sensors. Then in 1583 Galileo Galilei – in what was perhaps the first discovery in his illustrious scientific career – noticed a property of pendulums that revolutionized science and technology. While at mass in Pisa Cathedral, Galileo timed the swings of a large lamp that hung from the ceiling and which was set in motion as the verger lit the candles within it. Using his pulse as a timer, Galileo confirmed that no matter how hard the shove, or how wide the swing, the time of each swing – the period – was always constant. Twenty years later, Galileo returned to this subject and found that the mass of the bob also had no effect on the period of a pendulum, which meant that the swing time was defined by the length of the pendulum alone. He calculated that the period is proportional to the square root of the pendulum's length. (A pendulum with a period of one second is 99.4 millimetres [3.9 inches] long.) The first implication for this universal law was that swinging pendulums could be used as timekeepers, but more broadly it led to further research into the physics of oscillation, which helped scientists to understand waves, forces, gravity and even subatomic particles.

**CHRISTIAAN HUYGENS**
Galileo never made a clock using a pendulum. That first was achieved by Dutch scientist Christiaan Huygens (1629–1695) in 1656. Huygens was also an instrumental figure in several other fields, including optics, where he (correctly) suggested that light beams were oscillations, and astronomy: he discovered Saturn's ring system in 1659.

Observation of the motion of a lamp suspended from the ceiling of Pisa Cathedral led Galileo to a momentous discovery about the nature of movement.

LAWS OF MOTION **p.157** MEASURING TIME **p.186**

# Discovering Metabolism

c.1600

**SANTORIO SANTORIO:** *ARS DE STATICA MEDICINA* • PADUA, ITALY

**Key publications by Santorio Santorio**

*Methodi vitandorum errorum omnium qui in arte medica contingunt* 1602

*Commentaria in artem medicinalem Galeni* 1612

*Commentaria in primam Fen primi Canonis Avicennae* 1625

*Commentaria in primam sectionem Aphorismorum Hypocratis* 1629

The English word 'metabolism' is derived from the Greek word for change, and it refers to the thousands of chemical pathways that convert food into the substances needed to sustain life. However, this understanding, while intuitive to the modern mind, was not clear until the late nineteenth century. Ancient medics were aware that some of the material that entered the body was somehow lost, but they assumed that it evaporated away like an invisible sweat.

Santorio Santorio (1561–1636) investigated this issue with one of the longest-running experiments in science. He constructed a weighing chair so that he could record his weight before each meal and the weight of his food, and then he weighed his urine and faeces. (He also weighed himself before and after sleep, after sex and after working.) The data revealed that less material was excreted from the body than was ingested by it; on average he excreted 1.5 kilograms (3 pounds 5 ounces) for every 3.5 kilograms (7 pounds 11 ounces) he ate. This was the first clear evidence that material from the food was entering the body for use in fuelling and building the body.

**SANTORIO SANTORIO**

**Born in a part of what is now Slovenia but was then a province of the Venetian Republic, Santorio studied medicine at Padua. He began practising as a doctor in Venice in the 1580s and soon developed a reputation good enough to be tending to the city's wealthiest residents. In 1611 he returned to teach medicine in Padua.**

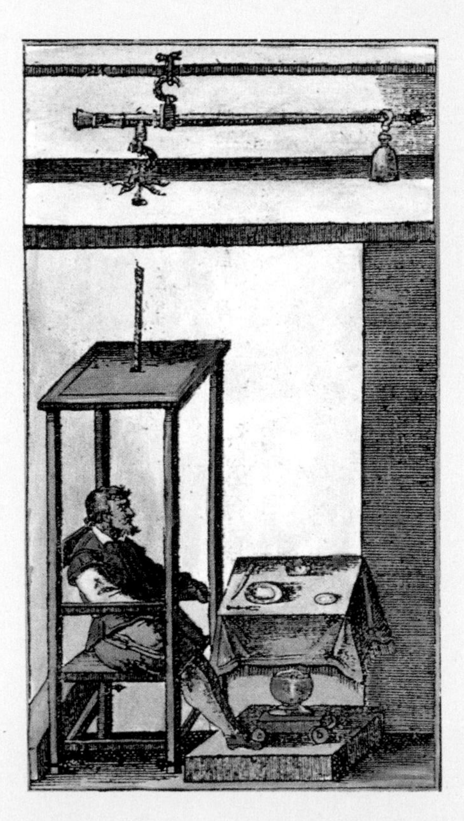

This seventeenth-century book illustration depicts Santorio weighing himself as part of the process of determining the difference between the amount of material his body ingested and excreted.

THE BIRTH OF MEDICINE **p.14** THE SCIENTIFIC REVOLUTION **p.18**

# Acceleration under Gravity

c.1600

**GALILEO GALILEI:** *ON MOTION* • PISA, ITALY

**Key publications by Galileo**

*The Starry Messenger* 1610

*Dialogue Concerning the Two Chief World Systems* 1632

*Discourses and Mathematical Demonstrations Relating to Two New Sciences* 1638

According to popular legend, Galileo hauled a pair of cannonballs, one small and one large, to the top of the Leaning Tower of Pisa and dropped them off the side of the building. He saw them 'fall evenly' and hit the ground at the same time. This was evidence that the weight of an object had no effect on how fast it falls; light things fall just as quickly as heavy ones. However, this event was entirely imagined by Galileo as a thought experiment. In real life, he rolled balls down a ramp fitted with equidistant bells, which chimed as each ball passed them. Galileo used this set-up to investigate the way bodies accelerate as they roll (or fall). He did this by comparing the distance travelled with the time taken, and alighted upon the law of fall, which states that the distance travelled is proportional to the square of the time taken. In other words, a ball falling for twice as long will fall four times further. To put it another way, a ball dropped from four times the height will only take twice as long to reach the ground.

**GALILEO GALILEI**

**Born in Pisa, the son of a musician, Galileo developed telescope astronomy and provided visual evidence that the Earth goes around the Sun. His work on motion paved the way for Newton and Einstein. Having grown up in precarious financial circumstances, Galileo cultivated noblemen to secure an income from his science and technology, but his work eventually led him into conflict with the Church, and he spent the last decade of his life under house arrest.**

Galileo accurately predicted that two cannonballs dropped from the Leaning Tower of Pisa would, notwithstanding their different sizes, hit the ground at the same time.

THE RENAISSANCE **p.17** THE SCIENTIFIC REVOLUTION **p.18**

# Circulation of Blood

**WILLIAM HARVEY:** *DE MOTU CORDIS* • FRANKFURT, GERMANY

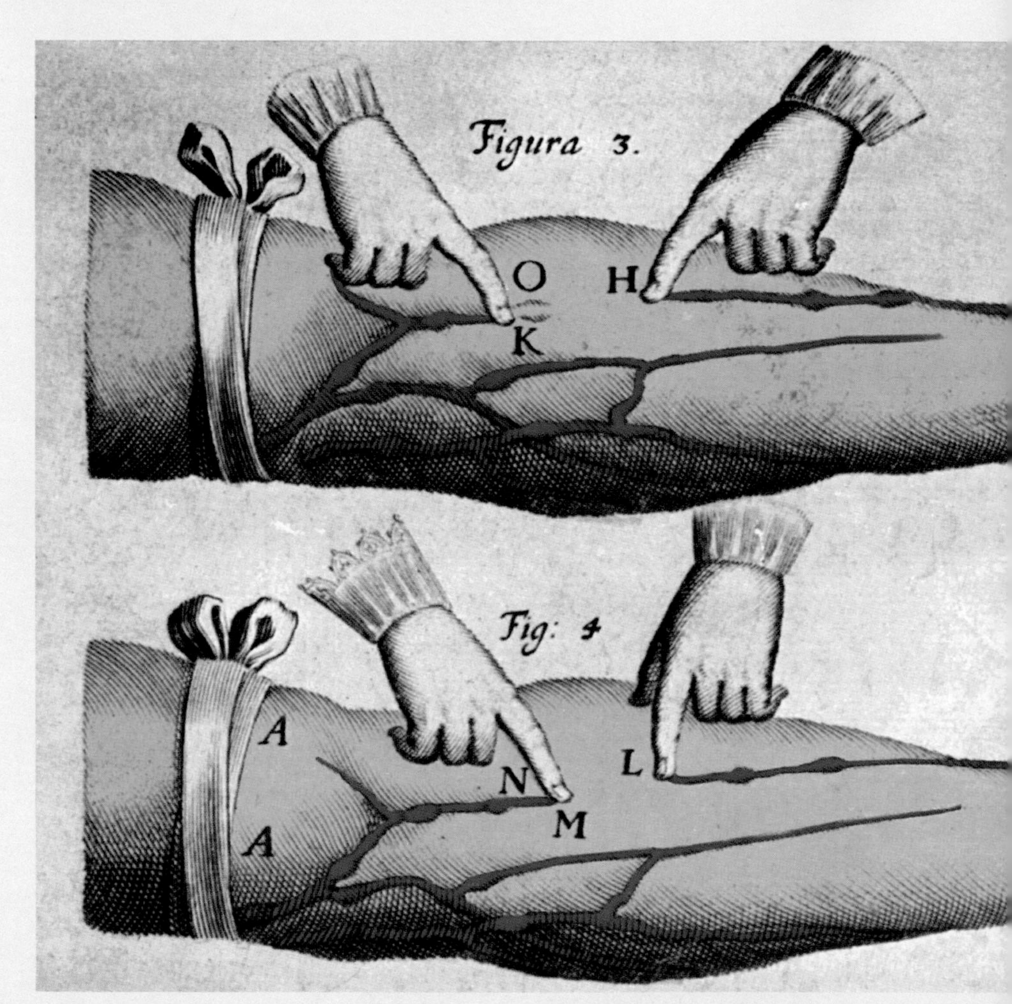


Harvey's 1628 textbook demonstrated that blood travels away from the heart through the arteries and returns to it through the veins, and that the flow is prevented from changing direction by a series of valves.

**Key publications by William Harvey**

***Exercitatio Anatomica de Motu Cordis et Sanguinis in Animalibus*** **(*An Anatomical Exercise on the Motion of the Heart and Blood in Living Beings*)** 1628

***Exercitationes de generatione animalium*** **(*On Animal Generation*)** 1651

THE BIRTH OF MEDICINE **p.14** GENETICS **p.31**

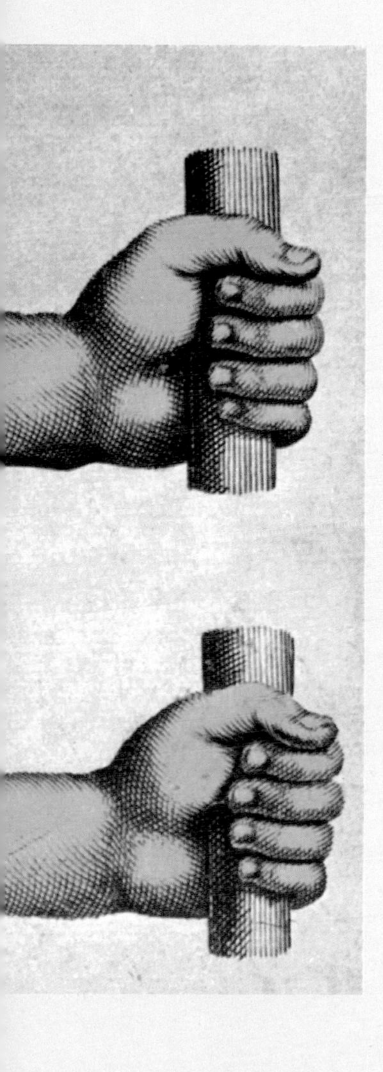

Until the seventeenth century, the body was believed to have two blood supplies: the veins carried dark blood distributing food from the liver, while the arteries carried air-filled blood from the lungs. In 1628, William Harvey presented a thorough rebuttal of this view in his book *An Anatomical Exercise on the Motion of the Heart and Blood in Living Beings* (*Exercitatio Anatomica de Motu Cordis et Sanguinis in Animalibus*). The dissection of human cadavers was not then permitted, so most of Harvey's insights came from vivisections, which frequently involved cutting open a range of animals to watch the living heart in action. Harvey addressed a prevailing theory that blood was made from food in the liver and constantly being used up by the body. He estimated that, if that was really the case, a human would have to make 250 kilograms (550 pounds) of fresh blood every day – far more than the weight of food eaten in that time.

The truth was that the veins and arteries held a constant volume of blood, and that the heart pumped it around in a loop. By tying off veins and arteries, Harvey demonstrated where blood entered and left the heart, and this showed that there are actually two loops of vessels. The right side of the heart receives oxygenated blood from the lungs and circulates it through the body. The left side collects the spent blood from the body and loops it through the lungs, from where it then feeds back to the right side of the heart.

**WILLIAM HARVEY**

**Born in Folkestone, England, William Harvey was a student of the surgeon Hieronymus Fabricius (1533–1619) at the medical school in Padua, Italy. It was Fabricius who discovered that veins have a one-way valve, a fact that later helped Harvey to puzzle out the circulation of blood. After graduating from Cambridge University in 1602, Harvey began an illustrious career in England, eventually becoming a court physician for the kings James I and Charles I before devoting more time to his anatomical research.**

LAWS OF THERMODYNAMICS **p.160** EVOLUTION BY NATURAL SELECTION **p.161**

# Weighing the Air

1643

**BLAISE PASCAL:** *TRAITÉ DE LA PESANTEUR DE LA MASSE DE L'AIR* • PARIS, FRANCE

Aristotle famously wrote that 'nature abhors a vacuum', by which he meant that a space could never contain nothing because something – usually air or water – would always rush in to fill the emptiness. In 1643, the Italian Evangelista Torricelli (1608–1647), a protégé of Galileo, investigated this by filling a tube with mercury, turning it upside down, and placing the open end in a bath of the same liquid metal. He found that the mercury did not flood out but always hovered at around the same height – 76 centimetres (30 inches) in modern units. What kept the mercury in place? Torricelli died before he could find out, and Frenchman Blaise Pascal (1623–1662) took up the subject. While some had suggested the empty gap at the top of the tube was a vacuum that pulled on the mercury, Pascal said it must be the downward push from the 'weight of the air'. If true, the mercury would drop with altitude because there was less air above pushing it down. In 1648, Pascal arranged for a Torricelli tube to be taken up a mountain and the height of mercury fell with every reading taken on the way to the summit. The 'weight of the air' is now better understood as air pressure, and is measured as the force applied by the atmosphere per unit of area. Torricelli's tube was the first barometer, a device for measuring pressure, and in Pascal's honour the scientific unit of pressure became the pascal (Pa).

**OTTO VON GUERICKE**

**German scientist Otto von Guericke (1602–1686) not only invented the electrostatic generator, which helped to set in train the study of electromagnetism, but also created an air pump that truly demonstrated the strength of air pressure. In 1663, he sucked the air from inside two iron hemispheres so that they were locked together by the surrounding air pressure. Two teams of eight horses hauling on both hemispheres could not prise them apart.**

THE BIRTH OF CHEMISTRY **p.20**

**Key publications by Blaise Pascal**

*Expériences nouvelles touchant le vide* 1647

*Traité du triangle arithmétique* 1665

*Pensées* 1670

This 1911 drawing by Max de Nansouty (1854–1913) depicts Florin Périer, at the behest of his brother-in-law Blaise Pascal, ascending the Puy de Dôme in France with a Torricelli tube containing mercury.

ATOMIC THEORY **p.159** SCIENTIFIC PROCESS **p.182**

# Gas Laws

1660s

**ROBERT BOYLE:** *NEW EXPERIMENTS PHYSICO-MECHANICALL: TOUCHING THE SPRING OF THE AIR, AND ITS EFFECTS* • LONDON, UK

**Key publications by Robert Boyle**
*The Sceptical Chymist* 1661
*Experimenta et Observationes Physicae* 1691

The idea of gas is relatively new. The word was not applied in a scientific sense until the end of the eighteenth century. The English word is derived from the Greek 'chaos' because, unlike a solid or liquid, gas lacks a fixed shape and volume. Before it was termed gas, this kind of material was simply known as air, and modern chemistry begins with the study of air carried out by Robert Boyle in the 1660s. Boyle principally used a pump to suck the air from jars and vessels, showing that a lack of air silenced the sound of a bell, snuffed out flames, and made plants and animals die. However, his scientific legacy is set as Boyle's law, which states that the pressure of a gas is inversely proportional to its volume. In other words, if a gas is squeezed into half its original volume, then the pressure it exerts will be doubled. Today this is one of three gas laws describing the behaviour of gas. The second is Charles's law from 1780. This states that a gas's volume is proportional to temperature, so a warming gas expands, a cooler one contracts. The third is Gay-Lussac's law from 1802, which states that the pressure of a gas is proportional to its temperature – heating a gas boosts its pressure. Armed with these laws, scientists were able to figure out the way gases (and all materials) are built from molecules and atoms.

**ROBERT BOYLE**
Born into a rich family of Irish landowners, Robert Boyle set out as a young man to enjoy the trappings of wealth. However, he soon turned to religion and devoted his material riches to the pursuit of learning. As now, the best scientific apparatus required the most precise engineering possible, and Boyle employed London's best glass-makers and metalworkers.
In his most famous book, *The Sceptical Chymist*, he meticulously demolished the claims of alchemy to be a science.

THE SCIENTIFIC REVOLUTION **p.18** THE BIRTH OF CHEMISTRY **p.20**

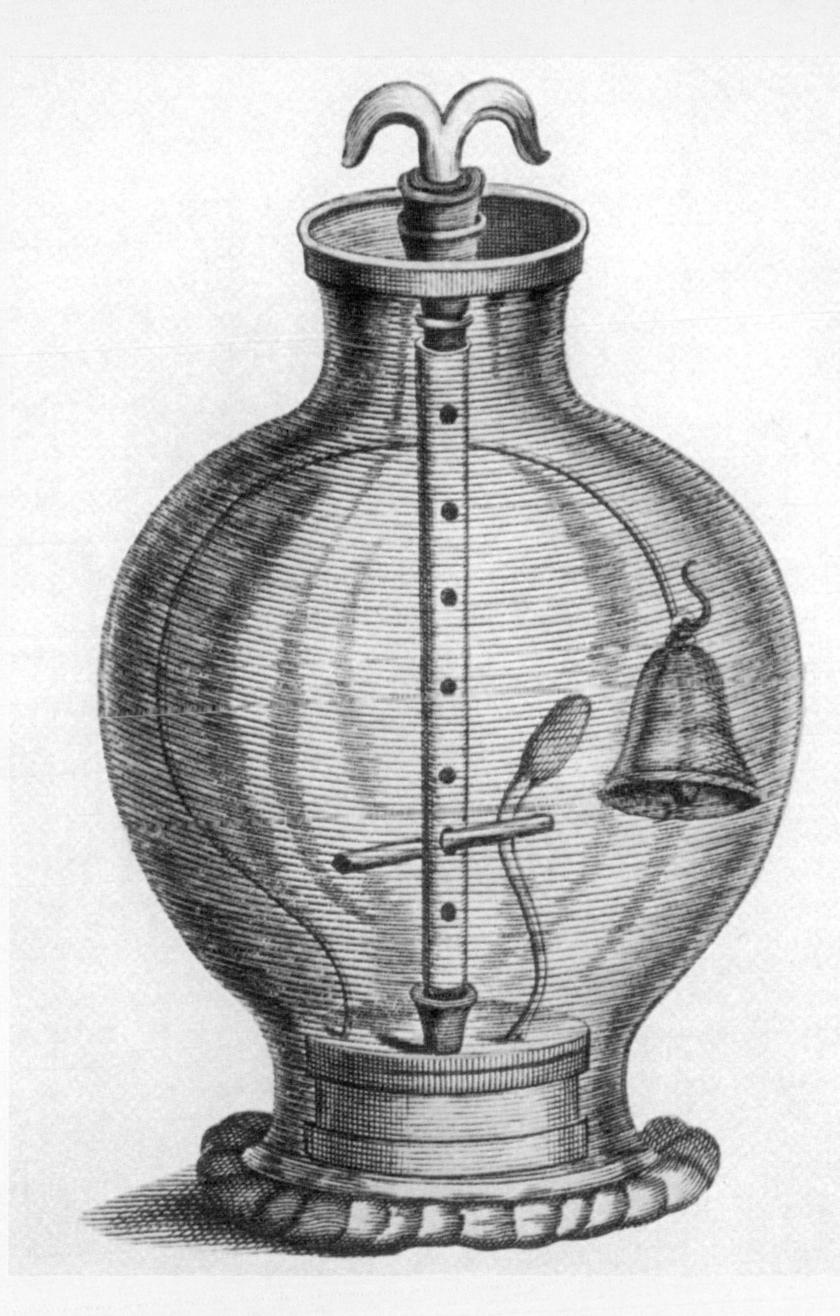

A drawing of the jar which Boyle sealed. He then pumped out the air in order to demonstrate that, in the absence of that gas, the bell inside could make no sound.

ATOMIC THEORY **p.159** LAWS OF THERMODYNAMICS **p.160** SCIENTIFIC PROCESS **p.182** THERMOMETERS **p.187**

# Hooke's Law

**ROBERT HOOKE:** *LECTURES DE POTENTIA RESTITUTIVA, OR OF SPRING. EXPLAINING THE POWER OF SPRINGING BODIES* • LONDON, UK

The view from inside the ground level of the Monument in London. Hooke used the staircase as a measuring scale.

THE RISE OF THE SCIENTIFIC INSTITUTION **p.19**

**Key publications by Robert Hooke**
***Some Considerations Touching the Usefulness of Experimental Natural Philosophy*** 1663
***New Experiments and Observations Touching Cold*** 1665

The Great Fire of London in 1666 is credited with ending the city's plague outbreak. It also offered several opportunities for the newly established Royal Society of London, the world's first academy of science, to show off the power of the scientific method. The society's secretary, Robert Hooke, was asked, along with London's great architect Christopher Wren (1632–1723), to build a monument to the fire. The result was a column – still known as the Monument – which Hooke designed to be a scientific instrument. Its hollow core and open roof were used as a telescope for making exact measurements of celestial objects; the spiral staircase inside was a scale for measuring how much materials stretched under a weight and then oscillated up and down. The results of these experiments led to Hooke's law, which states that the extension of a material is proportional to the weight (or other force) applied to it. Double the force, and the material stretches twice as much. The material resists the weight with a restoring force which is pulling it back to its original length. The same restoring force is present in a pendulum, pulling it back to the central start point, and Hooke's law helps to describe how these two forces are always working in opposition to create oscillations.

**ROBERT HOOKE**
**Born into lowly circumstances, Robert Hooke was often overshadowed (very unfairly) by his contemporaries, most notably Isaac Newton. The pair clashed over contributions to the theory of gravity, and Newton responded with his famous phrase: 'If I have seen further it is by standing on the shoulders of giants'. Historians disagree about whether Newton was acknowledging Hooke's contribution or making fun of his short stature.**

→ SCIENTIFIC PROCESS **p.182** MICROSCOPES **p.188** TELESCOPES **p.189**

# Discovery of Microorganisms

1682

**ANTONIE VAN LEEUWENHOEK:** *PART OF A LETTER FROM MR ANTONY VAN LEEUWENHOEK, CONCERNING THE WORMS IN SHEEPS LIVERS, GANTS AND ANIMALCULA IN THE EXCREMENTS OF FROGS* • DELFT, THE NETHERLANDS

**Key publication by Antonie van Leeuwenhoek**
*Observationes microscopicae Antonii Levvenhoeck* 1682

Although there is no undisputed date for the invention of the microscope, it is clear that by the 1620s the device, which uses a pair of lenses to magnify very small objects, was spreading across Europe, attracting attention as a scientific tool. In 1665, English scientist Robert Hooke compiled a lavishly illustrated account of his microscopy, and reported that a sliver of cork was made of small chambers – also known as cells – which were then recognized as the building blocks of living bodies.

In the 1670s, Antonie van Leeuwenhoek (1632–1723), a Dutch textile merchant, was struggling to confirm the number of threads – and so the quality – of his fabrics, so he assembled a hand-held microscope that magnified 300 times. When he then used this device to observe the natural world he found that what looked like clear pond water was full of microscopic life. He called the creatures '*dierkens*', which was translated into English as 'animalcules'. They even lived in raindrops. The objects which van Leeuwenhoek observed and had detailed drawings made of are now known to be bacteria and larger single-celled organisms such as amoebas and algae.

**ANTONIE VAN LEEUWENHOEK**
Born in Delft, van Leeuwenhoek was apprenticed as a draper and thrived in the textile business, opening his own shop at the age of 21. He subsequently created at least 500 hand-crafted lenses, and thus became renowned as one of the leading gentlemen scientists of the early modern age. In 1680 he was elected a fellow of the Royal Society of London, the premier science academy of the day.

THE SCIENTIFIC REVOLUTION **p.18** THE RISE OF THE SCIENTIFIC INSTITUTION **p.19**

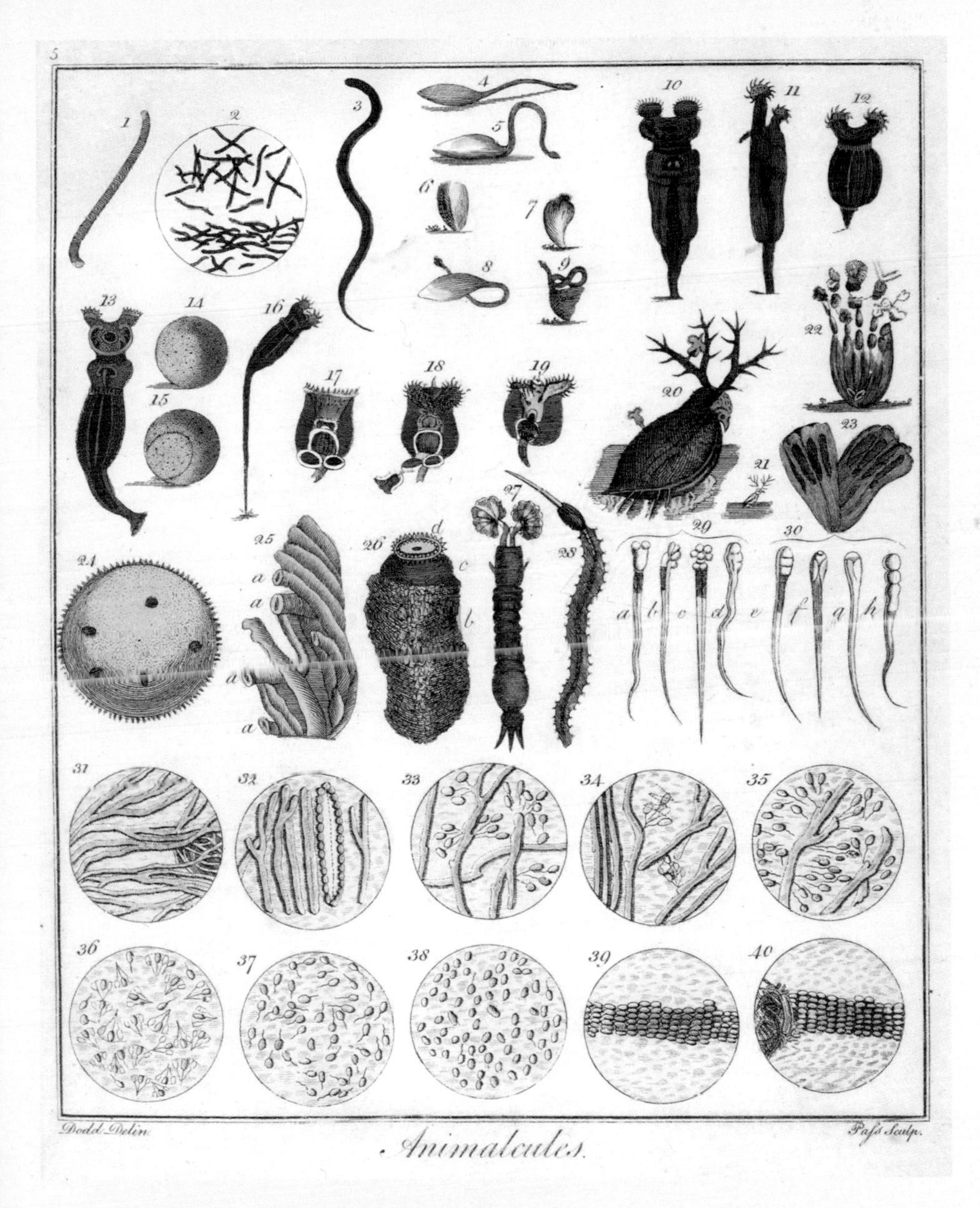


Some of the 'animalcules' observed by van Leeuwenhoek through his magnifying lens. His findings were derided at the time, but fully validated years later.

MICROSCOPES **p.188** TELESCOPES **p.189** CLADISTICS AND TAXONOMY **p.209**

# The Spectrum

1704

**ISAAC NEWTON:** *OPTICKS* • CAMBRIDGE, UK

In 1704, Isaac Newton, then the head of the Royal Society, published a book titled *Opticks*. As had been the case with Newton's more famous work on gravity, this book contained research carried out 30 years previously, this time into the nature of colour and light. Newton screened off his house in Cambridge University so that only a single shaft of sunlight entered it. The light was directed at a glass prism which dispersed the white sunlight into colours. This was a relatively new technique: although the rainbow of light produced was nothing new, it was Newton who described it as a spectrum, and it was he who determined that it contained seven colours: red, yellow, green, blue, indigo and violet. Despite being one of the most influential scientists in history, Newton was very superstitious. He thought having seven colours in the spectrum the most auspicious number, so much so that he more or less invented a new colour – indigo – which he named after a dark blue plant dye, a recent import to Britain from India. Newton's theory was that a light beam is made from particles, or corpuscles, which behaved like tiny balls, following his laws of motion; others suggested that light was a wave.

**ISAAC NEWTON**

**Newton is best known for his laws of motion, his universal theory of gravitation and for calculus, a mathematical technique for analyzing phenomena – including most natural ones – that are in constant change. He was also interested in heat and light, and famously invented a new design of telescope (best for astronomy) that used mirrors instead of expensive lenses. However, perhaps Newton's true passion was alchemy – he repeatedly tried to make gold – but of course he never succeeded in doing so.**

THE SCIENTIFIC REVOLUTION **p.18** THE RISE OF THE SCIENTIFIC INSTITUTION **p.19** 

This nineteenth-century wood engraving reimagines Newton dispersing white light through a prism.

**Key publications by Isaac Newton**

***De analysi per aequationes numero terminorum infinitas*** 1669

***Philosophiæ Naturalis Principia Mathematica*** 1687

***Scala graduum Caloris. Calorum Descriptiones & signa*** 1701

LAWS OF MOTION **p.157** UNIVERSAL GRAVITATION **p.158** MASS SPECTROMETRY **p.202**

# The Flying Boy

1730

**STEPHEN GRAY:** *A LETTER... CONTAINING SEVERAL EXPERIMENTS CONCERNING ELECTRICITY* • LONDON, UK

**Key publication by Stephen Gray**
*The Manuscript Letters of Stephen Gray, FRS*
1666/7–1736

Electricity has been studied since at least the sixth century BCE, when the philosopher Thales, often said to have been the first scientist, described how rubbing amber could create sparks and mysterious forces of attraction. (The word 'electric' comes from *elektra*, the Greek for amber.) Better methods of creating electricity were developed in the seventeenth century CE, including friction generators in which a glass ball was spun against a brush so that it was filled or charged with electricity. Nevertheless, electricity remained a static phenomenon. Then in 1730, Stephen Gray (1666–1736), a teacher, showed that a charge of electricity can flow, moving through some materials and blocked by others. At this time, electrical phenomena were often part of elaborate after-dinner shows in which 'electricians' performed for high-society guests. In keeping with this, Gray demonstrated his discovery in an equally theatrical way. A boy, one of his pupils, was suspended face down on silk ropes so that he hovered just above the floor, with his hands outstretched above dishes of gold leaf. Gray used a friction generator to charge the boy's feet and suddenly the gold leaf flew up through the air into the Flying Boy's hands, showing that the charge had been conducted through his body. Gray later found that metal and ivory also carried charge, while other materials, such as silk, blocked its passage. Today, the former materials are known as conductors, the latter as insulators.

**STEPHEN GRAY**
Born in Canterbury, England, Gray was apprenticed into his father's dye works. It was here he came across electrical phenomena, where materials often picked up charge as they were woven. Gray was then offered a position working for the Astronomer Royal, but later fell into poverty, and his friends arranged a job for him at the Charterhouse school, where he carried out his electrical experiments. In 1732 Gray received the first ever Copley Medal, the highest award given by the Royal Society.

THE RISE OF THE SCIENTIFIC INSTITUTION **p.19** ELECTRICITY **p.24**

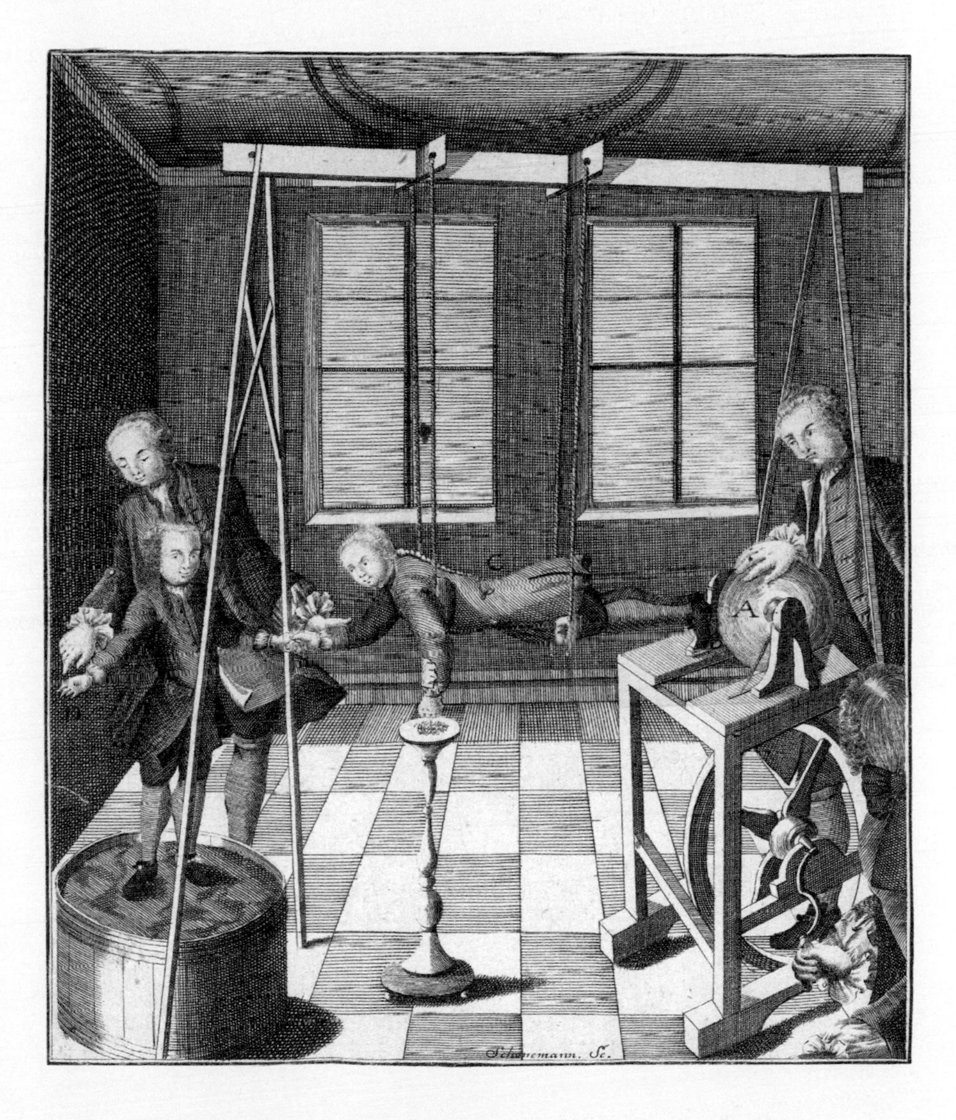

Stephen Gray's Flying Boy experiment accelerated the development of electricity from party trick to one of the most important amenities of human existence.

SCIENTIFIC PROCESS **p.182** CATHODE-RAY TUBE **p.193**

# Discovery of Photosynthesis

**JAN INGENHOUSZ:** *EXPERIMENTS UPON VEGETABLES, DISCOVERING THEIR GREAT POWER OF PURIFYING THE COMMON AIR IN THE SUN-SHINE AND OF INJURING IT IN THE SHADE AND AT NIGHT* • THORNHILL, YORKSHIRE, UK

In the mid-seventeenth century, Belgian alchemist Jan Baptist van Helmont (1580–1644) noted that, after planting a tree in a large pot, the weight of soil remained the same even though the tree grew steadily larger. He concluded that it must be the water added to the pot that supplied this growth and created the extra weight. This would explain why the tree would wilt when water was not provided. More than a century later, Jan Ingenhousz (1730–1799), a Dutch researcher, found that plants gave out oxygen during the day but took in carbon dioxide at night (as animals do all the time). Ingenhousz made this breakthrough while staying with his friend Joseph Priestley, who had recently discovered oxygen. Ingenhousz suggested that plants were taking, or 'fixing', carbon dioxide from the air and using it as a raw material for making wood and leaves as they grew. It took another century for a name to be given to this phenomenon: *photosynthesis*, which means 'making with light'. A green pigment named chlorophyll absorbs the red and blue wavelengths of sunlight so that energy is available to react carbon dioxide with water to make glucose, a simple sugar. Glucose made by photosynthesis is the foundation of almost all food chains on Earth.

NATURAL HISTORY AND BIOLOGY **p.22** ←

### JAN INGENHOUSZ

Ingenhousz was born in Breda, the Netherlands. After qualifying as a doctor, he moved to London, where he made a small fortune as an expert in variolation, a technique for inoculating against smallpox. He even provided the service to European royalty. As well as photosynthesis, he studied electricity and heat, and was in frequent correspondence with Benjamin Franklin and other leading researchers of the day.

Jan Ingenhousz and his servant Dominique demonstrate the previously unacknowledged capacity of vegetables to use light to generate energy.

**Key publications by Jan Ingenhousz**

*To Benjamin Franklin from Jan Ingenhousz* 15 November 1776

*Easy Methods of Measuring the Diminution of Bulk, Taking Place in the Mixture of Common Air and Nitrous Air...*1776

→ RADIOCARBON DATING p.197

# Oxygen

1774

**JOSEPH PRIESTLEY:** *EXPERIMENTS AND OBSERVATIONS ON DIFFERENT KINDS OF AIR* • BOWOOD, WILTSHIRE, UK

THE BIRTH OF CHEMISTRY **p.20**

Illustration of Priestley's experimental apparatus from his six-volume *Experiments and Observations on Different Kinds of Air* (1774–1786).

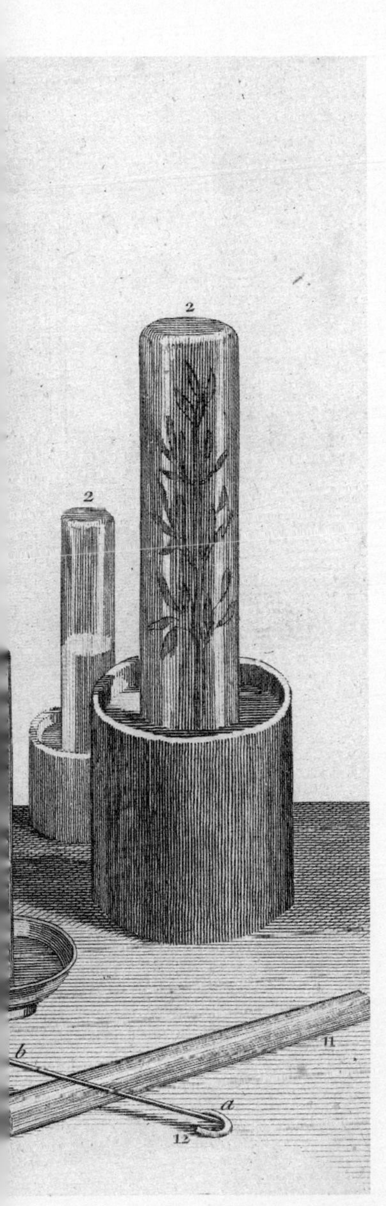

In 1756, Scotsman Joseph Black discovered 'fixed air', which seemed to be taken in, or fixed, by alkaline substances. (Today we call it carbon dioxide.) Fixed air snuffed out flames – and life itself – and so was a 'bad' air. Similarly, Englishman Daniel Rutherford revealed in 1772 that most of the air was 'bad' in this way when he discovered what was later named nitrogen. However, isolating the 'good' fraction of air, the part that fed flames and sustained life, was much harder. In the end it was discovered by chance in 1774 when another Englishman, Joseph Priestley, used lenses to focus sunlight and heat a sample of a lurid orange mercury mineral. The mineral broke down into pure mercury and gave off a colourless, odourless gas which did not snuff flames but instead made glowing embers burst into flames. Later the investigation of this new air was taken up by Antoine Lavoisier (1743–1794), the superstar of science at this time. The Frenchman showed that oxygen was involved in burning, producing ash, water and carbon dioxide and that animals – including humans – took in oxygen from the air and breathed out carbon dioxide. He also showed that water was a compound of oxygen and hydrogen, and he named both these gases: *hydrogen* means 'water-former'; *oxygen* means 'acid-former'.

**JOSEPH PRIESTLEY**

A Yorkshireman by birth, Priestley was a non-conformist clergyman and amateur scientist. In 1772 he showed how to make fizzy soda water by mixing in carbon dioxide, and this breakthrough beverage won him recognition enough to become the scientific advisor to the Earl of Shelburne, in whose house he discovered oxygen. Priestley was chased out of England in 1794 by nationalists who opposed his support for American independence, and he spent the rest of his life in Pennsylvania.

**Key publications by Joseph Priestley**

*The History and Present State of Electricity* 1767
*A New Chart of History* 1769
*Institutes of Natural and Revealed Religion* 1772–74
*Disquisitions relating to Matter and Spirit* 1777
*The Doctrine of Philosophical Necessity Illustrated* 1777
*Letters to a Philosophical Unbeliever* 1780

ATOMIC THEORY p.159 VALENCE BOND THEORY p.168

# Weight of the Earth

1789

**HENRY CAVENDISH:** *EXPERIMENTS TO DETERMINE THE DENSITY OF THE EARTH* • LONDON, UK

**Key publications by Henry Cavendish**
*Scientific Papers 1* 1921
*Scientific Papers 2* 1921
*The Electrical Researches of the Honourable Henry Cavendish* 1879

One of the big breakthroughs in the universal law of gravitation was the inverse square law, which showed how the force of gravity was inversely proportional to the distance between the objects. In other words, doubling the distance between two objects reduces the pull of gravity by a factor of four. But the law introduces a constant, G, known as 'big g', to calculate just how much force is in evidence between two masses. In 1789, Henry Cavendish wanted to measure the density and weight of planet Earth. To do that he needed a way of measuring G more precisely than ever before. He built a torsion balance, which was designed to twist slightly due to the pull of gravity between suspended large and small lead weights. Cavendish isolated the device in a building behind his London home and observed any movements from outside so as to not interfere with the forces in any way. Cavendish knew the masses of the weights and was able to calculate the size of the force between them from how much they moved. He arrived at a value for G, close to today's $6.67428 \times 10^{-11}$ (which is a very small number indeed). Knowing G, Cavendish calculated the mass of the Earth itself, based on the known acceleration due to gravity, and this told him that the planet's density is about five times greater than that of water.

**HENRY CAVENDISH**

As the third son of the Duke of Devonshire, Cavendish had almost unlimited resources to pursue his interests in science. However, at least in his early adulthood, he was painfully shy and built a separate staircase into his house so that he could access his lab without meeting staff. Cavendish became a leading figure of English science in 1766 when he discovered inflammable air, later known as hydrogen.

GEOLOGY AND THE EARTH SCIENCES **p.23**

Nineteenth-century drawing of Henry Cavendish.

UNIVERSAL GRAVITATION **p.158** STANDARD MEASUREMENTS **p.185**

# Conservation of Mass

1789

**ANTOINE LAVOISIER:** *TRAITÉ ÉLÉMENTAIRE DE CHIMIE* • PARIS, FRANCE

**ANTOINE LAVOISIER**
A Parisian by birth, Lavoisier trained as a lawyer but was always more interested in physics and chemistry and in 1768 he gained admission to France's prestigious Académie des sciences. He saw the French Revolution, which began in 1789, as an opportunity to reshape the state into a more equitable and inclusive form, but gradually the revolutionaries turned on people who had been rich before the social upheaval. Lavoisier came from the comfortably off middle class, and as a consequence he was guillotined in 1794.

The intuition that the world is constructed from a set of simple substances, or elements, proved to be correct. And there turned out to be a lot more than earth, air, fire and water, the four classical substances. The 'father of chemistry', Antoine Lavoisier drew up one of the first lists of elements in 1789 – although he still got a lot wrong. Nevertheless, he succeeded in debunking the ancient notion that materials slowly transmuted from one form to another, for example, water became earth. Lavoisier disproved this by boiling 1.36 kilograms (3 pounds) of water in a sealed flask for 100 days, after which there was still the same weight of water. This discovery established the principle that matter could not be destroyed or created but was constantly combining and recombining into compound substances. Most famously, Lavoisier showed that water was created by burning hydrogen (or flammable air) with oxygen – and was also able to split into these constituents again. As a wealthy French aristocrat, Lavoisier had access to expensive, precision-built apparatus which enabled him to prove that the materials before a chemical reaction weighed precisely the same as the products thereby produced, including any gases given off.

**Key publications by Antoine Lavoisier**
*Réflexions sur le phlogistique* 1783
*Traité élémentaire de chimie* 1789
*Mémoire sur la chaleur* 1780

THE BIRTH OF CHEMISTRY **p.20**

A celebrated 1788 portrait of Lavoisier and his wife, Marie-Anne Pierrette Paulze, by Jacques-Louis David (1748–1825).

THE PERIODIC TABLE **p.162** VALENCE BOND THEORY **p.168** DISTILLATION **p.204**

# Animal Electricity

1791

**LUIGI GALVANI:** *DE VIRIBUS ELECTRICITATIS IN MOTU MUSCULARI COMMENTARIUS* • BOLOGNA, ITALY

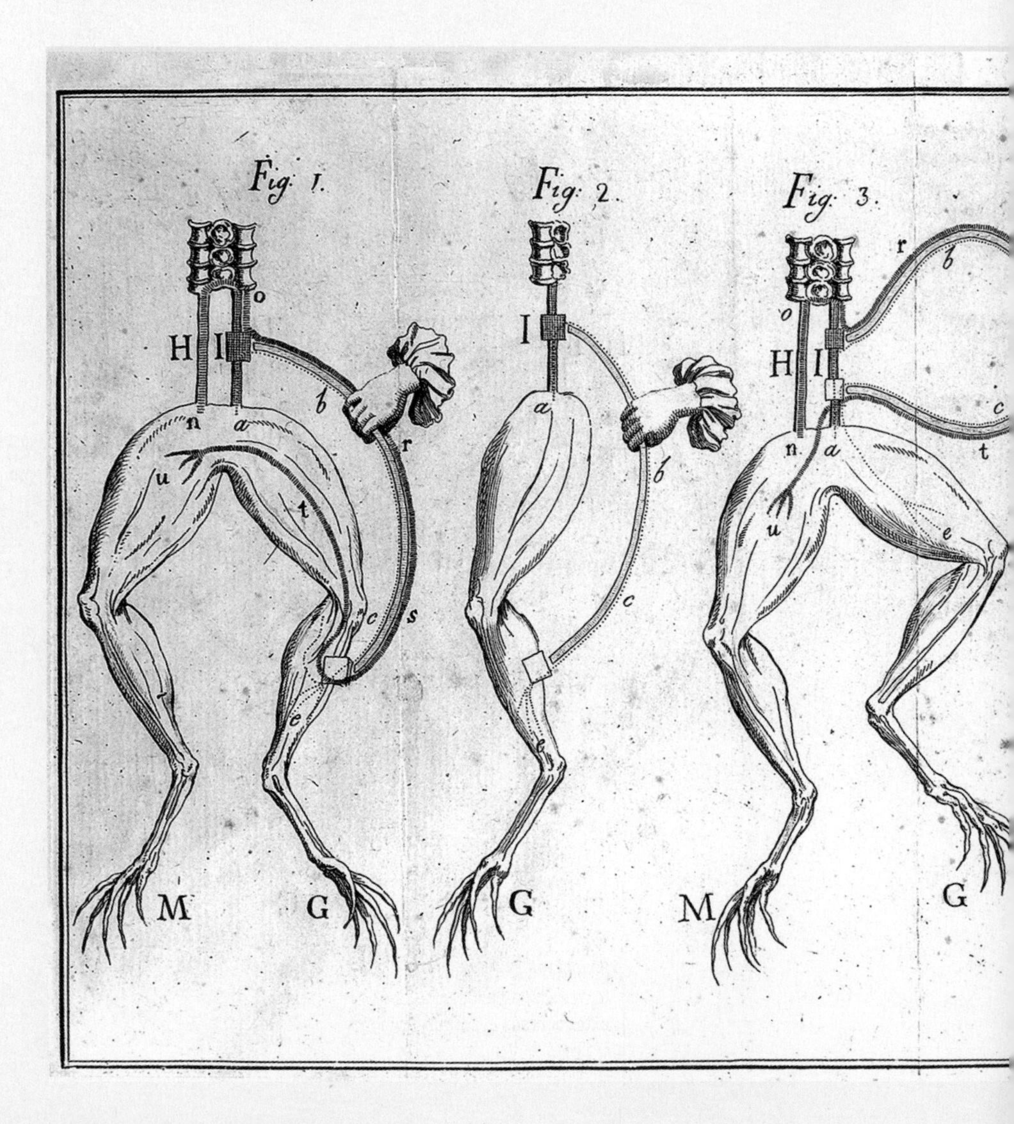


1793 diagram of the experiment Galvani performed on the dismembered body of a frog, showing the positioning of the metals on the amphibian's sciatic nerve.

SCIENCE AND THE INDUSTRIAL REVOLUTION **p.21** ELECTRICITY **p.24**

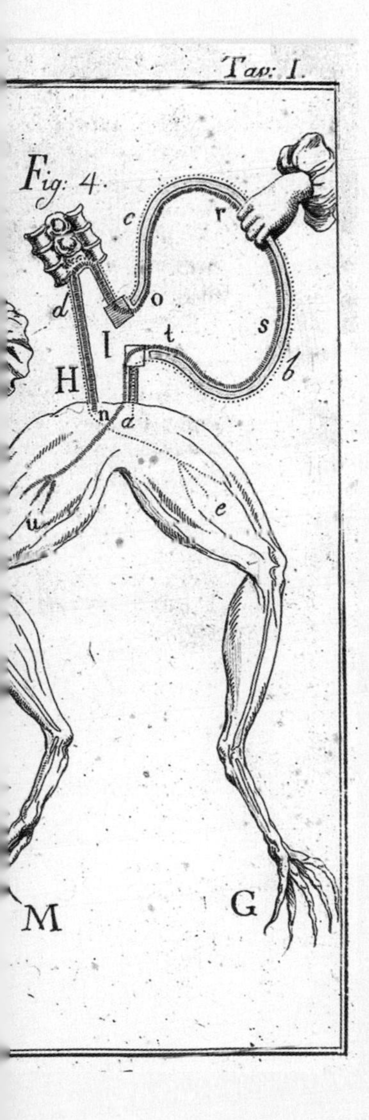


The first evidence that electricity could be made to run in a circuit and put to useful work came from an unlikely source: frogs' legs. The discovery was accidental and made when Luigi Galvani (1737–1798) – an expert in anatomy, not physics – dangled some severed frogs' legs on a copper hook placed on an iron railing. Spookily, the legs began to twitch as if they were alive. (In another version of this story, the effect was seen when the legs were held in place for dissection with metal pins that were touching a metal knife.) Either way, Galvani noted the significance of the two different metals and created an arc-shaped connector with one end made of copper and the other of iron. When he touched the tips – points that are now known as electrodes – to both ends of a frog's leg he saw sparks and the characteristic twitching. He was able to recreate this effect with all kinds of dismembered appendages. Giovanni Aldini (1762–1834), Galvani's nephew, travelled the world using this technique to reanimate the corpses of recently executed murderers, making them gurn – reports of which are said to have inspired Mary Shelley's *Frankenstein*. Galvani proposed that he had discovered a vital electric force that powered animals. While it is true that muscles and nerves use electric pulses, what was really happening was that the two metals and salty animal fluids created a primitive battery.

**LUIGI GALVANI**

Galvani spent his whole life in Bologna, Italy. He considered joining the clergy before opting to train in medicine, and took extra courses in surgery, which led him to become a professor of anatomy at Bologna University. (Elevation to this post was helped by the fact that the previous incumbent was his father-in-law.)

**Key publication by Luigi Galvani**

Giovanni Aldini, *An account of the late improvements in galvanism* 1804

FOUR FUNDAMENTAL FORCES **p.105**

# Vaccination

**EDWARD JENNER:** *AN INQUIRY INTO THE CAUSES AND EFFECTS OF THE VARIOLÆ VACCINÆ, A DISEASE DISCOVERED IN SOME OF THE WESTERN COUNTIES OF ENGLAND, PARTICULARLY GLOUCESTERSHIRE, AND KNOWN BY THE NAME OF THE COW POX* • LONDON, UK

In this satirical etching of 1802 by James Gillray, the patients being vaccinated by Edward Jenner are developing the features of cows.

THE BIRTH OF MEDICINE **p.14**

**EDWARD JENNER**

**The son of a Gloucestershire vicar, Jenner was himself variolated as a child, an experience that led to health problems throughout his life. At the age of 14 he was apprenticed to a surgeon and completed his medical training in London. He returned to Gloucestershire to work as a local doctor. His work on vaccination won him many plaudits, and in 1821 he was appointed as one of King George IV's doctors.**

In 1796, English doctor Edward Jenner (1749–1823) performed the first vaccination to make his patient, an eight-year-old boy named James Phipps, immune to smallpox, a deadly and disfiguring disease. The procedure involved injecting the patient with a serum containing cowpox, a disease common among dairy workers that was similar to but much less deadly than smallpox. Jenner coined the term 'vaccination' from *vacca*, the Latin for 'cow'. Jenner's breakthrough did not come out of the blue. In the 1720s, the practice of variolation had arrived in Britain. This involved collecting pus from the blisters of a smallpox sufferer and smearing it into a small cut made in the skin. This ancient practice had been shown to create immunity – if it didn't kill you first, which was a distinct possibility. By the 1770s, several researchers had realized that people who contracted cowpox never seemed to catch smallpox. A Dorset farmer named Benjamin Jesty is reported to have injected his family with cowpox to protect them during a 1774 smallpox outbreak. Jenner, who served another farming community in Gloucestershire, studied this procedure for 20 years before being sure it was safe. A few days after treating James, Jenner then injected him with a sample of smallpox pus, and the boy suffered no illness.

**Key publications by Edward Jenner**

***Further Observations on the Variolæ Vaccinæ, or Cow-Pox*** 1799

***A Continuation of Facts and Observations relative to the Variolæ Vaccinæ, or Cow Pox*** 1800

***The Origin of the Vaccine Inoculation*** 1801

CLINICAL TRIALS **p.208**

# Proving Extinction

1796

**GEORGES CUVIER:** *ESSAY ON THE THEORY OF THE EARTH* • PARIS, FRANCE

**Key publications by Georges Cuvier**

*Essais sur la géographie minéralogique des environs de Paris* 1811

*Le Règne Animal* 1817

*Théorie de la terre* 1821

NATURAL HISTORY AND BIOLOGY **p.22** GEOLOGY AND THE EARTH SCIENCES **p.23**

Fossils have been interpreted in several ways throughout history. The ancient Chinese thought that they were dragon bones; spiral-shaped ammonites were seen as serpent-stones made of the body of a coiled snake. Some suggested that fossils were placed in the rocks by God to test faith in His creation, while others simply assumed that they were the remains of life from long ago and that they belonged to the same kinds of animals as those that currently roamed the Earth.

In 1796, French zoologist Georges Cuvier (1769–1832) found that the fossil remains of American elephants were significantly different from today's Asian and African elephants. He thus revealed the startling fact that animals that lived in the past – as seen in the fossil record – were not the same species as those that lived in his own time. The idea that species became extinct (died out completely) rapidly became a standard feature of theories of evolution. The fossil record also gave widespread credence to the notion that Earth was millions of years older than conventional religious teachings of the era would acknowledge.

**GEORGES CUVIER**

**Although he is known as the founding father of palaeontology, Cuvier did not believe that extinct fossils represented the ancestors of today's wildlife and strongly opposed theories of evolution. Instead, it was his view that extinct creatures were evidence of successive great floods, including the one involving Noah.**

Georges Cuvier lecturing on palaeontology at the Muséum national d'Histoire naturelle, Paris.

EVOLUTION BY NATURAL SELECTION **p.161** ENDOSYMBIOSIS **p.173** RADIOCARBON DATING **p.197**

# Electrolysis

c.1800

**HUMPHRY DAVY:** *THE COLLECTED WORKS OF SIR HUMPHRY DAVY*
LONDON, UK

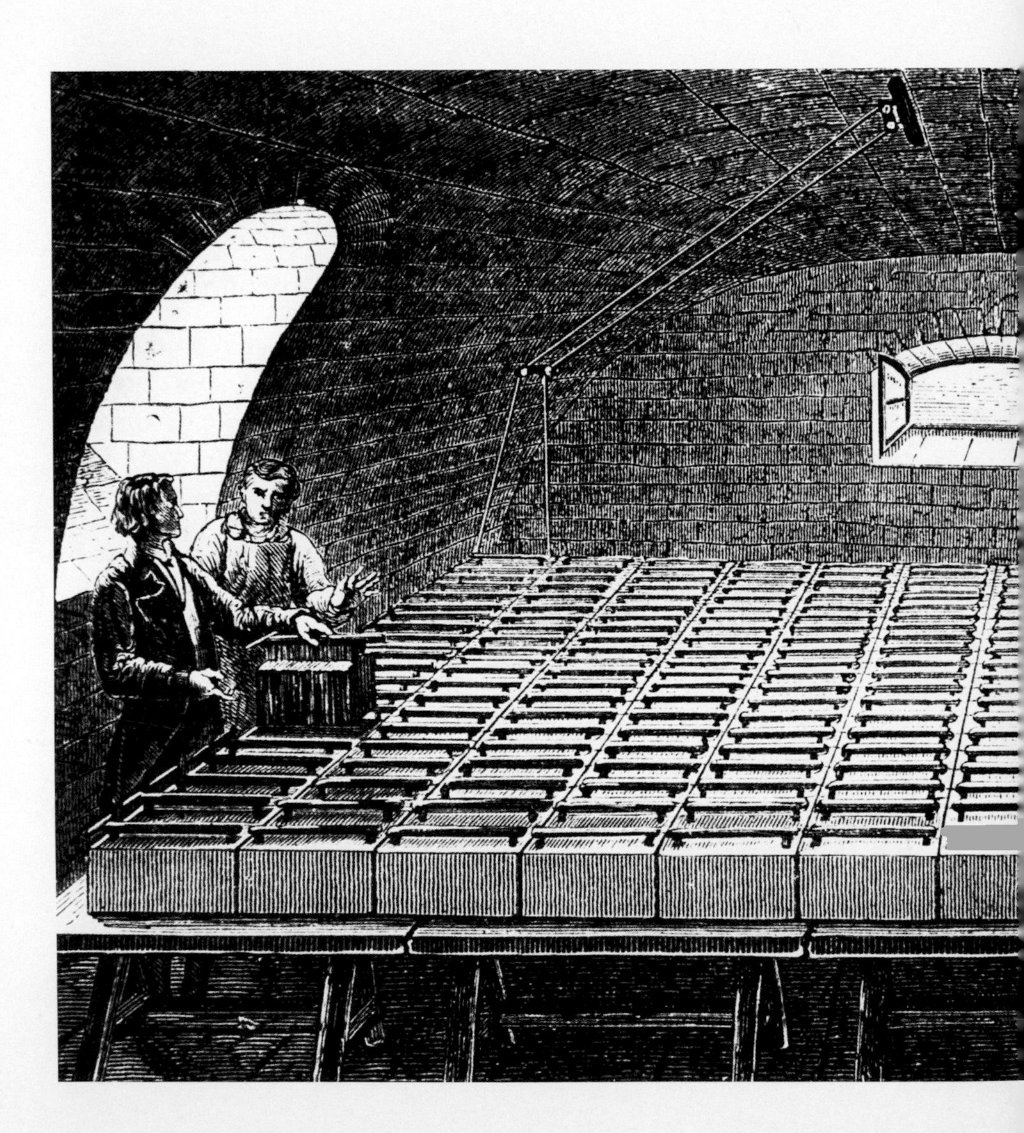

ELECTRICITY **p.24**

**Key publications by Humphry Davy**

*Researches, Chemical and Philosophical; Chiefly Concerning Nitrous Oxide, or Dephlogisticated Nitrous Air, and Its Respiration* 1800

*Elements of Chemical Philosophy* 1812

*The Papers of Sir H. Davy* 1816

The electric battery was invented in 1800 and scientists wasted no time in using this new source of electrical current as a tool for investigating nature. The most successful of them was Humphry Davy (1778–1829), a Cornish chemist who had achieved fame as the inventor of a safety lamp for miners and for discovering laughing gas. Davy was by now one of the chief researchers at the Royal Institution in London, where probably the world's largest battery was being created in the basement. Davy knew that an electric current could split water into hydrogen and oxygen gas, a process later named electrolysis, and so he used the Institution's vast battery to see what happened when solutions of dissolved minerals were electrified. The first two substances he used were potash and soda ash, both thought to be elements at the time. Electrolysis revealed that they were actually compounds of two unknown metals, which Davy named potassium and sodium. The same technique later enabled Davy to claim the discovery of magnesium, calcium, boron, strontium and barium; he also showed that chlorine and iodine were elements. Electrolysis gave a strong hint that the atoms of constituent elements were held together within these compounds by some kind of electric force.

**HUMPHRY DAVY**

**The son of a Penzance woodcarver, Humphry Davy's education was paid for by his godfather, John Tonkin, who then had him apprenticed as an apothecary. However, his employers soon complained about the teenager's dangerous chemical experiments. Davy then moved to Bristol to work at the Pneumatic Institute, a gas research centre, where he helped to discover nitrous oxide (laughing gas), and his career was set. He became a patron of science and apprenticed one Michael Faraday, but the pair fell out in the 1820s over the latter's secret research into electric motors.**

A nineteenth-century engraving depicting the giant battery built in the basement of the Royal Institution's building in London's Mayfair district.

THE PERIODIC TABLE **p.162**

# Double Slit Experiment

1804

**THOMAS YOUNG:** *EXPERIMENTS AND CALCULATIONS RELATIVE TO PHYSICAL OPTICS* • LONDON, UK

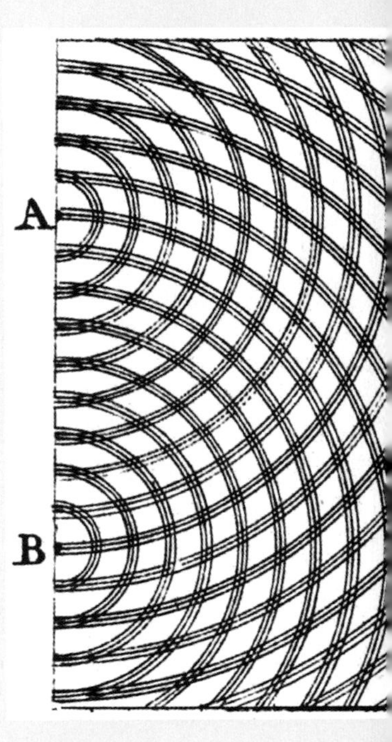


Until the nineteenth century there were two competing and mutually exclusive theories of light. Newton had proposed that light was a cascade of particles, while Huygens contended that beams were made of waves. Huygens was better able to describe the behaviour of light by treating it as a wave. For example, light diffracts (spreads out in all directions) when it meets a narrow gap. The same thing happens to ripples propagating across a pond, creating a half-circle of wavefronts after passing through a gap. It was harder to see how light particles could do this, but nevertheless many remained wedded to Newton's ideas. In 1804 Thomas Young (1773–1829) decided to compare light with water waves, specifically looking at the phenomenon of interference, where waves combine. If the waves are in sync with each other, the new wave is twice as big as the original wave. Out-of-sync waves cancel each other out. Young demonstrated this by passing water ripples through two slits. The wavefronts formed on the far side then interfered with each other to create a distinctive pattern of calm water and taller ripples. Then Young shone a light through two narrow slits, and the beam created a distinctive pattern of bright and dark bands, where the light waves had combined and cancelled out just like the ripples. This was the proof: light is a wave, not a stream of particles.

**THOMAS YOUNG**

**From a wealthy English West Country Quaker family, Young practised as a London doctor until 1801 when he began a career in academic research. As well as studying light, Young is remembered for Young's modulus, which is a way of measuring the linear elasticity of materials.**

THE NEW PHYSICS **p.27**

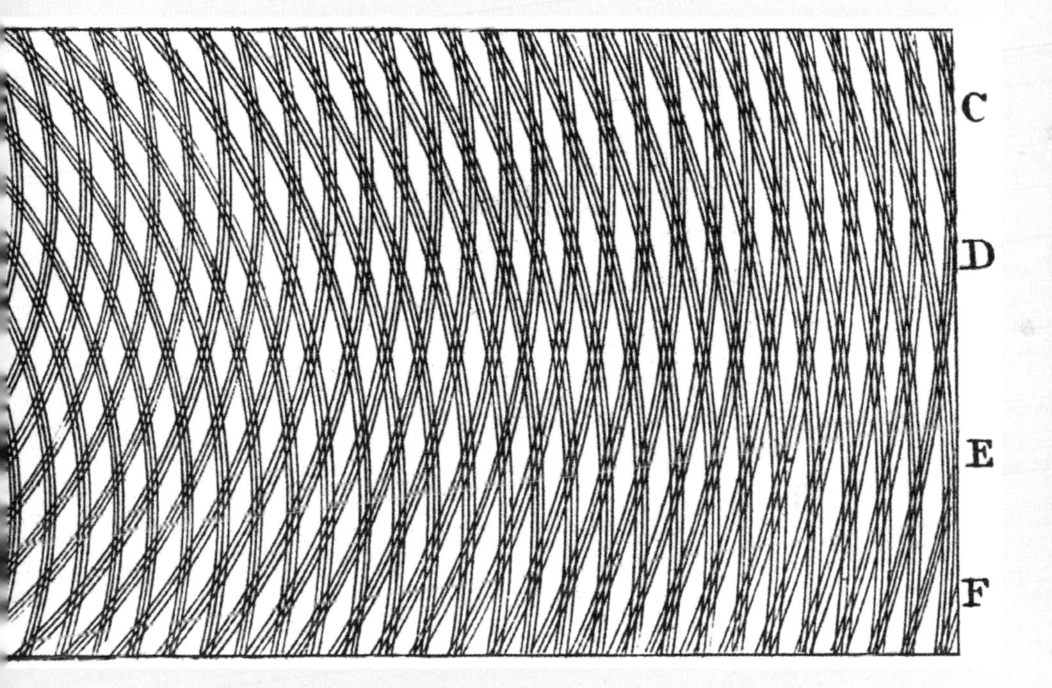


Young's sketch of two-slit diffraction. The narrow slits at A and B are the sources, and waves interfering at various phases are at C, D, E and F.

**Key publications by Thomas Young**

*A Course of Lectures on Natural Philosophy and the Mechanical Arts* 1807

*Miscellaneous Works of the Late Thomas Young, M.D., F.R.S.* 1855

QUANTUM PHYSICS **p.167** LASERS **p.195**

# Electromagnetic Unification

1820

**HANS CHRISTIAN ØRSTED:** *EXPERIMENTA CIRCA EFFECTUM CONFLICTUS ELECTRICI IN ACUM MAGNETICAM* • COPENHAGEN, DENMARK

**Key publication by Hans Christian Ørsted**
*Aanden i Naturen* 1850

**ANDRÉ-MARIE AMPÈRE**
The fact that an electric current forms a magnetic field is Ørsted's law, named after its discoverer, but most early work on electromagnetism was done by André-Marie Ampère (1775–1836), who quantified the relationship between electric current and magnetic forces. The unit of current, the amp (A), is named for him.

The forces of magnetism by which iron and other metals attract and repel each other have been known about since at least 600 BCE, when Greek natural philosophers first wrote about them. Within 500 years of that, Chinese geomancers began using magnetic compasses in their rituals, and over the following centuries the compass became a world-changing navigation tool. However, the phenomenon of magnetism was treated separately from that of electricity, despite both creating attractive and repulsive forces summed up as the shared property of 'opposites attract and likes repel'. In 1820, a science demonstration by Hans Christian Ørsted (1777–1851), a Danish professor, changed all that, simultaneously merging both phenomena into a single field of electromagnetism and laying the foundations for modern technology. Ørsted's lecture in Copenhagen had intended to demonstrate that electricity made a wire glow and give out heat. However, a compass for use in a later demonstration sat beside the wire on the bench, and Ørsted noticed that when the wire was electrified, the compass swung to point at it, and then realigned to the north when the current was switched off. The implication was clear – electrified objects create a magnetic field – and the possible applications of this effect, from electric motors to telecommunications, became the focus of inventors for decades to come.

This nineteenth-century engraving shows Ørsted discovering that a pivoted magnetic needle turns at right angles to a conductor carrying an electric current.

ELECTRICITY **p.24**

# Carnot Cycle

1824

**SADI CARNOT:** *REFLECTIONS ON THE MOTIVE POWER OF FIRE* • PARIS, FRANCE

**Key publication by Sadi Carnot**
*Reflections on the Motive Power of Heat* 1890

By the dawn of the nineteenth century, the Age of Steam was in full spate. The powerful steam-powered engines installed in factories across Europe were evidence enough that they worked, but quite how was still not fully understood. A young French soldier, Sadi Carnot (1796–1832), decided to figure out how a steam engine converted heat into motion. He concluded that all such engines work by heat moving from a source (the fuel) to a cold sink outside the engine. The work done by the engine was equivalent to the temperature drop between the source and the sink, and the energy transfer happened in four stages now known as the Carnot cycle. In stage one the heat makes the steam (or any gas) expand without changing its temperature. In stage two, the expansion continues, but the gas cools as its heat is converted to a pushing motion on the engine's piston. In stage three, the gas is compressed by the falling piston but does not warm up. Instead, heat energy escapes into the sink. In stage four, further compression heats the gas as it returns to its original state.

**SADI CARNOT**
**Christened Nicolas Léonard Sadi Carnot, Carnot's unusual name was taken from a Persian poet admired by his father, Lazare. Carnot Senior was a revolutionary leader who was exiled once the French monarchy was restored in 1814, and his soldier son found his military career going nowhere. Sadi devoted more time to steam engines. In 1832 he died of cholera.**

Sadi Carnot in 1813 at the age of 17 in the traditional uniform of a student of the École Polytechnique, painted by Louis-Léopold Boilly (1761–1845).

SCIENCE AND THE INDUSTRIAL REVOLUTION **p.21**

# Brownian Motion

1827

**ROBERT BROWN:** *A BRIEF ACCOUNT OF MICROSCOPICAL OBSERVATIONS MADE ON THE PARTICLES CONTAINED IN THE POLLEN OF PLANTS*
LONDON, UK

**ROBERT BROWN**
Born and raised in Montrose, Scotland, Brown dropped out of Edinburgh medical school to join the army. Fortunate to barely see action, Brown spent deployments collecting botanical specimens and had a special interest in using the microscope to examine plants. In the 1800s, Brown took up a position as a scientist on a voyage to Australia, which propelled him into a full-time career as a botanist.

The principle that substances are all composed of tiny particles known as atoms, often arranged in clusters or molecules, was largely established by the 1820s, and a few decades later the constant vibration or motion of these minute physical entities was being used to understand the way heat and other forms of energy moved through and between materials. No matter how robust these theories – and they remain undisputed today – they were all based on abstracted evidence. Then in 1905 Albert Einstein pointed out that visual proof had been available all along in the work of a Scottish botanist's investigations of pollen in 1827. The botanist was Robert Brown (1773–1858) who, while examining under a microscope the pollen of ragged robin, a pink flower collected from the Pacific Northwest, saw the pollen grains expel tiny particles. Now known to be grains of starch and oil droplets, these tiny specks danced randomly before his eyes. Brown recreated the effect with inanimate coal dust, but was none the wiser about what he was seeing. It was left to Einstein to clarify the matter: the random movement of particles – what had become known as 'Brownian motion' – was due to collisions between visible molecules and invisible molecules in the water. Here was confirmation that atomic theories were correct.

**Key publications by Robert Brown**
*Prodromus florae Novae Hollandiae et Insulae van Diemen* 1810
*On the natural order of plants called Proteaceae* 1810
*List of new and rare plants, collected during the years 1805 and 1810, arranged according to the Linnaean system* 1814

THE NEW PHYSICS **p.27** THE SIZE OF THE UNIVERSE **p.28** 

Robert Brown at the age of 82.

ATOMIC THEORY p.159 RELATIVITY p.163 MICROSCOPES p.188 BUBBLE CHAMBERS p.198

# Vitalism

1828

**FRIEDRICH WÖHLER:** *ÜBER KÜNSTLICHE BILDUNG DES HARNSTOFFS*
BERLIN, GERMANY

**Key publications by Friedrich Wöhler**
*Lehrbuch der Chemie* 1825
*Grundriss der Anorganischen Chemie* 1830
*Grundriss der Organischen Chemie* 1840

For centuries the start point for biologists was that living things were distinct from inanimate materials because they were run though with a 'vital force' – some kind of imperceptible material that made it possible for the conditions of life to exist. Those special conditions were needed for the complex 'organic' chemicals of life to be created outside the everyday chemistry that could be investigated in the lab or observed happening elsewhere in the natural world. If proof were needed, any non-vital physical or chemical change irreversibly 'denatured' the biochemicals into materials that lacked the vital force and were thus unable to support life.

This ancient theory was exploded by a single accidental discovery. In 1828 German chemist Friedrich Wöhler (1800–1882) was in his lab attempting to make ammonium cyanate, but found he had produced urea instead. Urea was a relatively simple and well-known organic substance found in urine. According to the theory of vitalism, urea could only be made using an animal's kidney. This discovery was proof that the chemistry of life worked in exactly the same way as inorganic chemistry. Wöhler's experiment became the foundation of organic chemistry, which studies all carbon-based compounds. Later the study of chemicals involved in life was hived off as biochemistry.

**FRIEDRICH WÖHLER**
**Wöhler was the son of a veterinarian, but he chose to pursue a medical career, graduating in 1823. He then moved quickly to become a full-time chemist and began work with Jöns Jacob Berzelius (1779–1848), an influential Swedish chemist who was later instrumental in promoting Wöhler's negation of vitalism. In later life, Wöhler was the first to purify the metals beryllium and yttrium.**

CELL THEORY **p.25**

This portrait of Friedrich Wöhler was painted around 1896 by Konrad von Kardorff (1877–1945).

CENTRAL DOGMA OF BIOLOGY **p.172** RADIOCARBON DATING **p.197**

# Doppler Effect

1842

**CHRISTIAN DOPPLER:** *ÜBER DAS FARBIGE LICHT DER DOPPELSTERNE UND EINIGER ANDERER GESTIRNE DES HIMMELS* • PRAGUE, CZECHIA

**Key publication by Christian Doppler**
*Über das farbige Licht der Doppelsterne* 1842

In 1842, Austrian physicist Christian Doppler (1803–1853) had a question about waves, such as those in sound or light. Every wave has a wavelength, which is the distance from the peak of one wave to the peak of the next. The wavelength is locked to the wave's frequency, which is a measure of how many wavelengths are completed per second. A short wavelength produces a high frequency. The ears perceive the frequency of a sound wave as pitch, with high-pitched sounds having high frequencies. Light perception is similar. High-frequency light is blue, low-frequency light is red, with yellows and greens in between. So Doppler wondered what happened to frequencies when the source of the wave was moving relative to the observer. He imagined a boat moving into waves. The crests would hit the bow more frequently – as if the wavelengths had been shortened – than if the vessel were stationary. Doppler proposed that the same thing happened with starlight, so when a star moves towards the viewer, the wavelengths of its light are compressed so that its colours shift towards blue in the spectrum. A star moving away is 'redshifted'. The so-called Doppler effect is more apparent in the siren of a speeding emergency vehicle. The sound is compressed to a high pitch as it approaches, and then lengthened to a deeper tone as it passes and moves away.

**CHRISTIAN DOPPLER**
Born in Salzburg, Doppler moved to Vienna to study philosophy and mathematics at the university, where he was admitted as an assistant professor at the age of 26. In 1835 he moved to Prague, and it was there that he presented the work that made his name before delving into a wide range of problems in mathematics and physics. In 1848 he moved back to Vienna and became head of the Institute for Experimental Physics, but died soon after from lung disease.

THE SIZE OF THE UNIVERSE p.28

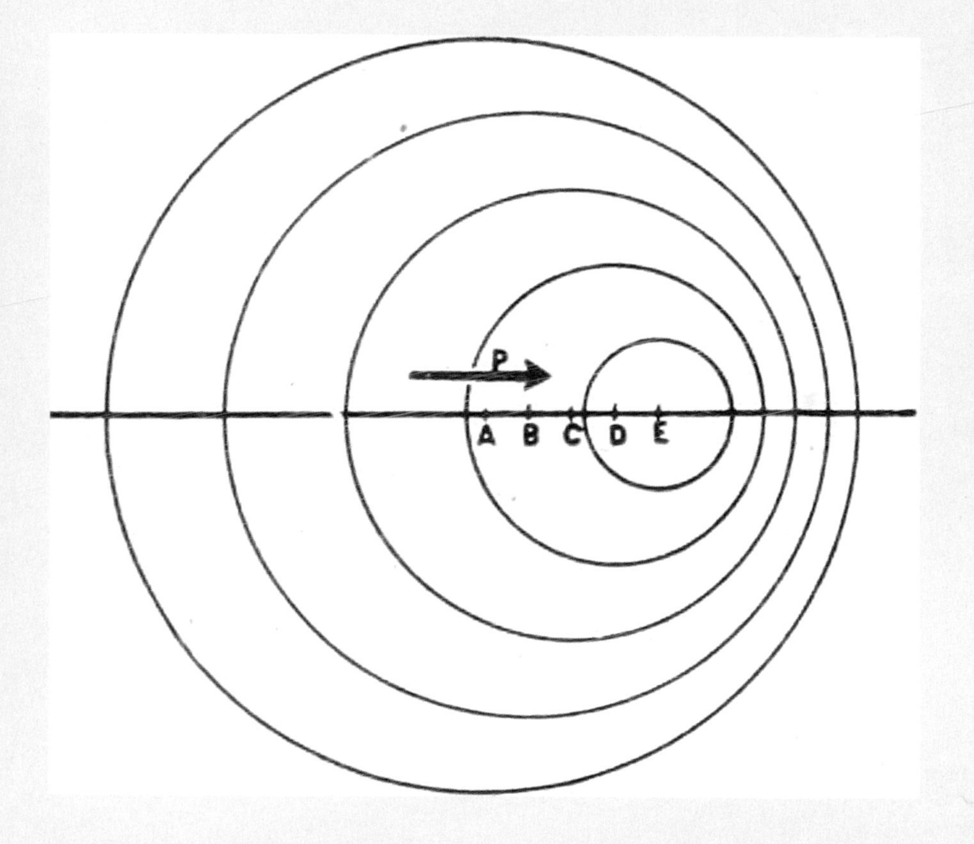


Diagram illustrating how perceived sound changes as its source advances towards, then passes and recedes away from the listener along path P.

BIG BANG **p.169** TELESCOPES **p.189**

# Mechanical Equivalent of Heat

1843

**JAMES PRESCOTT JOULE:** *THE MECHANICAL EQUIVALENT OF HEAT*
SALFORD, UK

**BENJAMIN THOMPSON, COUNT RUMFORD**
Before Joule's famous experiment, Thompson (1753–1814) had demonstrated the same phenomenon in 1798. A Loyalist American in the Revolutionary War, Thompson married Antoine Lavoisier's widow and founded the Royal Institution in London. While working as a military engineer in Bavaria, Thompson showed that grinding a blunt drill into a cannonball immersed in a barrel of water made the water warm up – and after two and a half hours start to boil. However, Thompson's experiment lacked the empiricism of Joule's and was hence not accepted by the scientific community.

In the 1840s the background understanding of what would become the first law of thermodynamics – energy is not destroyed or created, but transformed from one type to another – was taking shape. The electricity motor and generator showed that electrical energy could become motion energy and vice versa. German scientist Julius von Mayer (1814–1878) proposed a link between heat and motion, noting that rough seas were warmer than calm ones. In 1843 English brewer James Prescott Joule (1818–1889) became interested in the link between energy and heat while investigating the efficacy of electrical heating systems for use in his beer-making. He built an apparatus to measure the mechanical equivalent of heat, which involved a pulley system in which a falling 0.45 kilogram (one pound) weight spun a stirrer inside a sealed tank of water. Joule wanted to know how far the weight needed to fall in order to raise the temperature of one pound of water by one degree Fahrenheit. After running the experiment for several hours, he reached an answer: the weight had to fall 255 metres (838 feet) to achieve that level of heating. This showed that a small amount of motion energy of the weight and stirrer was becoming heat energy. Joule is reported to have attempted to measure this effect in a Swiss waterfall while on honeymoon, comparing the water temperature before and after it falls, but only succeeded in getting wet. He and others would later explain that heat energy itself was actually the motion of atoms and molecules within a substance.

SCIENCE AND THE INDUSTRIAL REVOLUTION **p.21** 

**Key publications by James Prescott Joule**

*On the Heat evolved by Metallic Conductors of Electricity, and in the Cells of a Battery during Electrolysis* 1841

*On the Changes of Temperature Produced by the Rarefaction and Condensation of Air* 1844

Nineteenth-century coloured wood engraving showing Joule at work on his researches into the mechanical equivalent of heat.

LAWS OF THERMODYNAMICS **p.160** SCIENTIFIC PROCESS **p.182** THERMOMETERS **p.187**

# Speed of Light

1849

**HIPPOLYTE FIZEAU:** *SUR L'EXPÉRIENCE RELATIVE À LA VITESSE COMPARATIVE DE LA LUMIÈRE DANS L'AIR ET DANS L'EAU* • PARIS, FRANCE

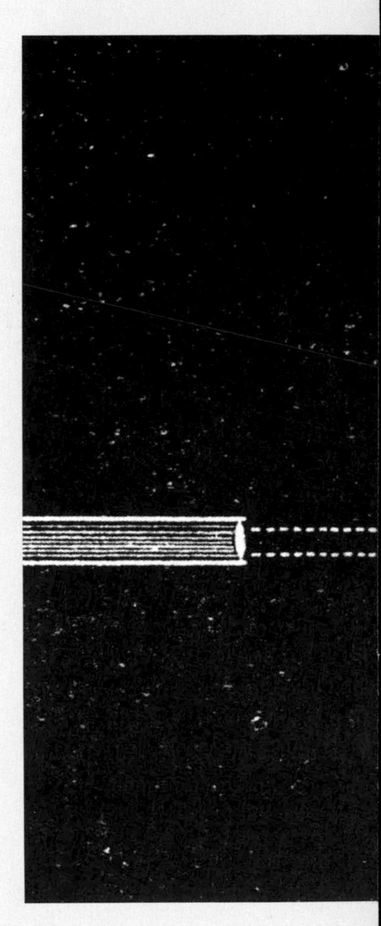

Even the most cursory investigation shows that the speed of light is very fast indeed, too fast to measure with a clock. Some suggested that light speed was infinite, and that the Sun illuminated the whole Universe in an instant. However, in 1676 Danish astronomer Ole Rømer (1644–1710) demonstrated that light from outer space becomes visible at different times in different parts of Earth, thus proving that light travels at finite speed.

The first apparatus capable of accurately measuring the speed of light was developed in 1849 by Frenchman Hippolyte Fizeau (1819–1896), and improved upon by Léon Foucault (1819–1868). It used a telescope lens to focus a beam of light into a second telescope 8 kilometres (5 miles) away, which promptly reflected the beam straight back so that it made a 16-kilometre (10-mile) round trip. At the start of its journey, the beam met a spinning cog and shone through the gaps in the wheel's teeth. Thus the beam had a precise on-off flicker. The scientists adjusted the speed of the cog until the returning beam was blocked by the next cog tooth, thus preventing the light returning to its start point. Initially, Fizeau used this method to calculate a speed of 315,000 kilometres per second, which was a bit high. Foucault upgraded the experiment in 1862, and arrived at a figure of 298,000 kilometres per second, which is within 0.6 per cent of the accepted modern value of 299,792 kilometres per second (670,616,629 miles per hour).

**OLE RØMER**

**While working at the Paris Observatory in 1676, Rømer tracked the orbit of Io, a moon of Jupiter, which regularly appeared and disappeared behind the giant planet for periods of time that varied according to the relative proximity of Jupiter to observers on Earth. The time elapsed between eclipses of Io becomes shorter as Earth moves closer to Jupiter and longer as Earth and Jupiter draw further apart. From this Rømer calculated the speed of light as 212,000 km/s.**

THE BIRTH OF CHEMISTRY **p.20**

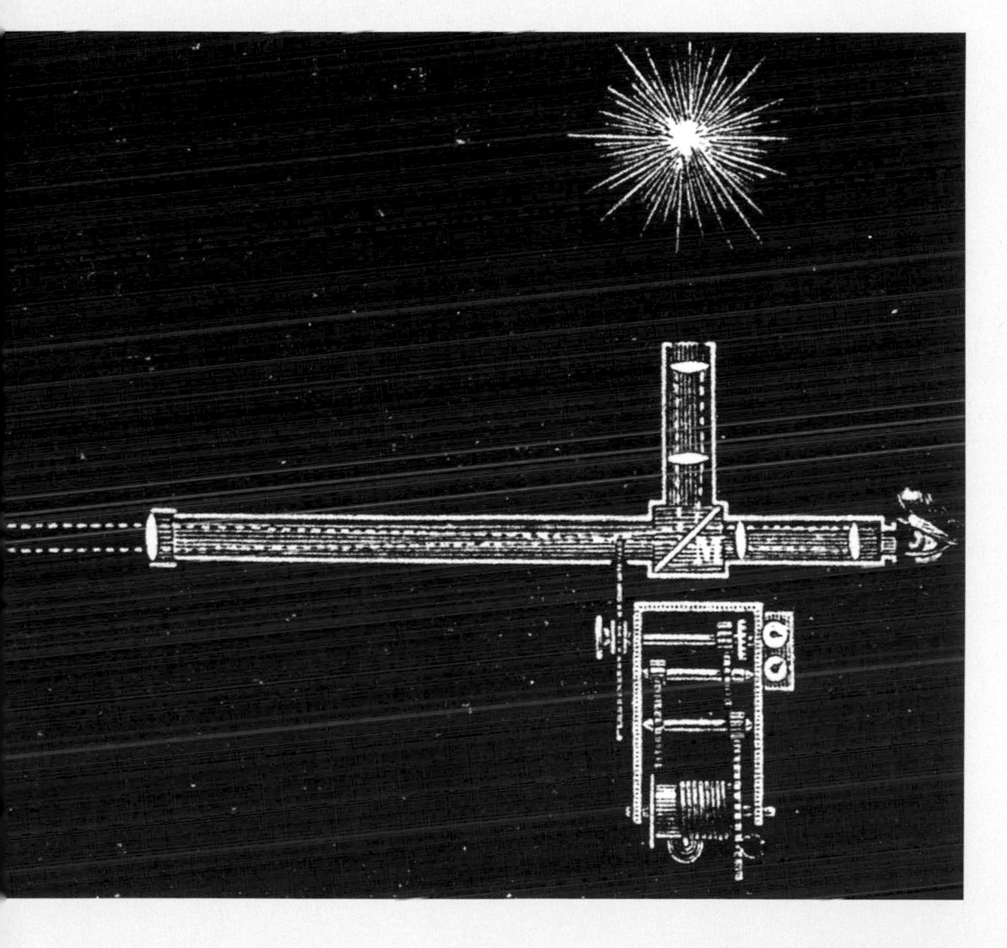


Engraving depicting Hippolyte Fizeau's apparatus for determining the speed of light.

RELATIVITY p.103 MEASURING TIME p.186 PARTICLE ACCELERATORS p.199

# Rotation of Earth

1851

**LÉON FOUCAULT:** *RECUEIL DES TRAVAUX SCIENTIFIQUES* • PARIS, FRANCE

**Key publication by Léon Foucault**
*Recueil des travaux scientifiques de Léon Foucault*
1878

On his deathbed in 1543, Nicolaus Copernicus announced the fact – then incredibly heretical – that Earth moved around the Sun, and the apparent motion of the Sun through the sky was caused by Earth rotating on an axis once a day. It took another century or so for astronomers and the wider science world to accept this view – and centuries more for other authorities – but there was no direct evidence of Earth's rotation until 1851, when Léon Foucault erected a large pendulum, 67 metres (220 feet) long, in Paris's Panthéon. As Galileo had explained long before, when a pendulum swings it always moves back and forth in the same plane. And when Foucault set his immense pendulum swinging it appeared to do just that. Beneath the pendulum, Foucault laid out a bed of fine sand and positioned the pendulum so that a point extending from the bob just touched the sand, leaving grooves as it swung. As the hours passed, the marks left by the pendulum appeared to shift in a clockwise direction. Eventually it had turned through 360 degrees and was back swinging in its original position. Nevertheless, Galileo's pendulum law was not incorrect, and the pendulum's swing had not shifted direction. Instead, the whole world had been turning beneath it.

**LÉON FOUCAULT**
As a child, Foucault was educated largely in his Paris home, and seemed destined for medical school, but a phobia of blood led him to switch to physics. He took an interest in light, working on ways of measuring its intensity, and investigated the nature of invisible infrared heat rays. Although the rotation experiment had originally been conceived in the 1600s, the apparatus is now known as a Foucault pendulum.

ANCIENT ASTRONOMERS **p.12**

Foucault's pendulum still stands in the Panthéon in Paris.

LAWS OF MOTION **p.157**

# Spectroscopy

1859

**GUSTAV KIRCHHOFF:** *ABHANDLUNGEN ÜBER EMISSION UND ABSORPTION*
HEIDELBERG, GERMANY

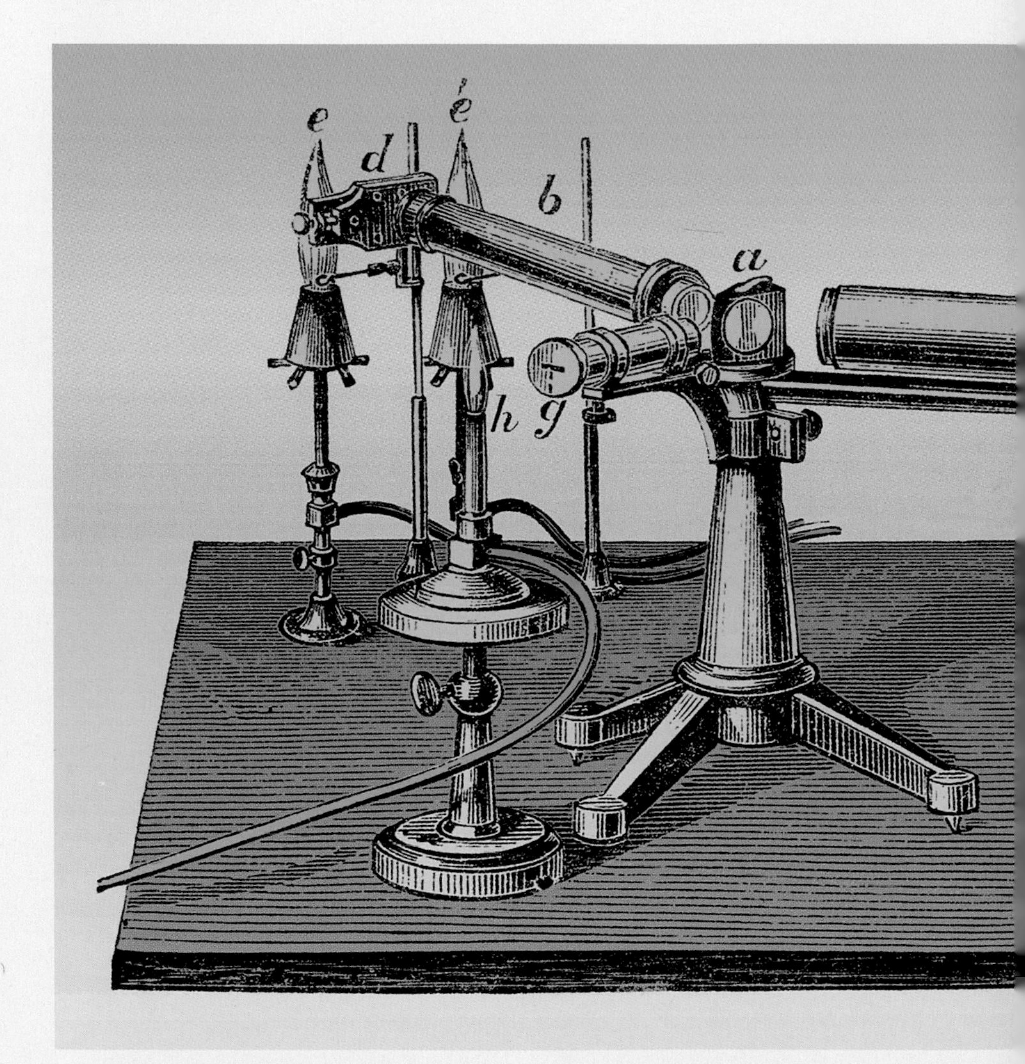


Bunsen-Kirchhoff spectroscopic apparatus, 1869.

THE SIZE OF THE UNIVERSE **p.28**

**Key publications by Gustav Kirchhoff**
*Vorlesungen über mathematische Physik* 1876–1894
*Gesammelte Abhandlungen* 1882

The presence of some elements, mainly metallic ones, can be determined by setting fire to a specimen and observing the colour of its flame.

Alchemists used this flame test centuries before 1859, when two German scientists, Robert Bunsen (1811–1899) and Gustav Kirchhoff (1824–1887), advanced the process to create a powerful means of identifying elements. Bunsen is famed for his laboratory gas burner, which is designed to bring a clean source of heat that does not interfere with the chemical reactions of burning. Then, through the use of a spectroscope (a device that uses a prism to split the light from the flames into its constituent colours), Kirchhoff formulated the three laws of spectroscopy: 1) Hot objects produce a full spectrum of colours (as does the Sun); 2) A hot gas, such as that inside a flame, emits only a small set of specific colours, and every element has its own unique emission spectrum; and 3) A cold gas absorbs colours from the full spectrum, creating a unique absorption spectrum, which is seen as dark gaps.

Bunsen and Kirchhoff recorded the spectra of all known elements and also discovered two new metals, rubidium and caesium. Modern astronomers use these spectra to identify the materials in stars, nebulae and other deep-space objects.

**JOSEPH VON FRAUNHOFER**

**The dark lines of absorption spectra seen in starlight are named Fraunhofer lines in honour of Joseph von Fraunhofer (1787–1826), the German optical scientist who invented a way of making lenses that did not distort colours. Born into poverty and orphaned at 11, Fraunhofer was nearly killed in 1801 when the workshop where he was apprenticed as a glassmaker collapsed. Maximilian Joseph, the Prince of Bavaria, oversaw the rescue and took the boy into his care. Fraunhofer invented the spectroscope in 1814.**

QUANTUM PHYSICS **p.167** BIG BANG **p.169** TELESCOPES **p.189** MASS SPECTROMETRY **p.202**

# Germ Theory

1861

**LOUIS PASTEUR:** *LES MICROBES ORGANISÉS, LEUR RÔLE DANS LA FERMENTATION, LA PUTRÉFACTION ET LA CONTAGION* • PARIS, FRANCE

A famous experiment carried out by Frenchman Louis Pasteur (1822–1895) in 1861 proved that disease and decay were caused by microorganisms in the air. This germ theory of disease went against the prevailing view at the time, which was that germs arose spontaneously from rotting material rather than causing the rot. Pasteur had been tasked in the 1850s with figuring out why wine spoiled. He revealed that the culprit was microscopic yeasts that chemically altered the drink.

Pasteur knew that bacteria grow in open containers of meat broth. He also knew that if the broth is boiled for an hour in a sealed container, no bacteria will grow in it, and moreover that bacteria are found in airborne dust. Based on this knowledge, Pasteur put broth into several sealed flasks, then boiled the contents to kill any pre-existent bacteria. Once the broths were sterilized, Pasteur broke off some but not all of the seals: the broth in those that were left intact developed no bacteria; the contents of those that were broken developed microbial life.

This germ theory was further backed up in the 1870s as specific bacteria were shown to be the causes of particular diseases.

**Key publications by Louis Pasteur**
*Études sur le vin* (*Studies on Wine*) 1866
*Traitement de la Rage* (*Treatment for Rabies*) 1886

**LOUIS PASTEUR**
Louis Pasteur was a chemist, not a medical doctor, and even before his breakthrough over germ theory he had already entered the history books with his discovery of chirality. This is the idea that complex molecules can exist in two forms, each a mirror image of the other. His later career was devoted to developing vaccines for diseases such as anthrax and rabies, and of course pasteurization, a flash-heating process that kills germs in milk without altering the flavour.

CELL THEORY **p.25** PUBLIC HEALTH **p.26** SCIENCE AND THE PUBLIC GOOD **p.29**

This 1885 painting by Finnish artist Albert Edelfelt (1854–1905) shows Louis Pasteur in his laboratory. Note the swan-necked flask, a crucial part of his bacteria experiments.

MICROSCOPES **p.188** CLINICAL TRIALS **p.208**

# Existence of Genes

1865

**GREGOR MENDEL:** *VERSUCHE ÜBER PFLANZENHYBRIDEN* • BRNO, CZECHIA

**Key publication by Gregor Mendel**
*Versuche über Pflanzenhybriden (Experiments in Plant Hybridization)* 1866

The knowledge that offspring inherit characteristics from their parents has been used for millennia by animal and plant breeders to create domestic breeds and varieties that embody the most desirable features of their forebears.

Yet the process of inheritance is not a simple blending of characteristics from both parents. Although inheritance was a central plank of Darwin's 1859 theory of evolution by natural selection, neither its proponent nor anyone else really understood how it worked. However, an Austrian monk named Gregor Mendel (1822–1884) was already conducting exhaustive research into the way various attributes of pea plants were passed from generation to generation. Mendel's seven years of experimentation involved carefully controlling which plants bred with which. In 1865 he presented his findings, one of which was that all plants (and by extension all sexually reproducing organisms) have two copies of each gene, but only one of them is passed on to its offspring, and the transmission of each gene is independent of the others. Finally, Mendel described a law of dominance, according to which certain inherited traits, such as the tallness of a plant, are dominant and will always be seen when inherited. Recessive traits, such as shortness, remain hidden unless the organism inherits two versions of this gene.

**GREGOR MENDEL**
Born in what is now part of Czechia, Mendel had an insecure childhood as his family struggled to grow enough food to eat. He became a monk to ensure he could pursue his interests without that kind of worry. Initially his genetic investigations centred on breeding honeybees, but he created a very aggressive variety that had to be destroyed, so he turned his attention to pea plants. His ground-breaking work was overlooked until the start of the twentieth century.

CELL THEORY **p.25** GENETICS **p.31**

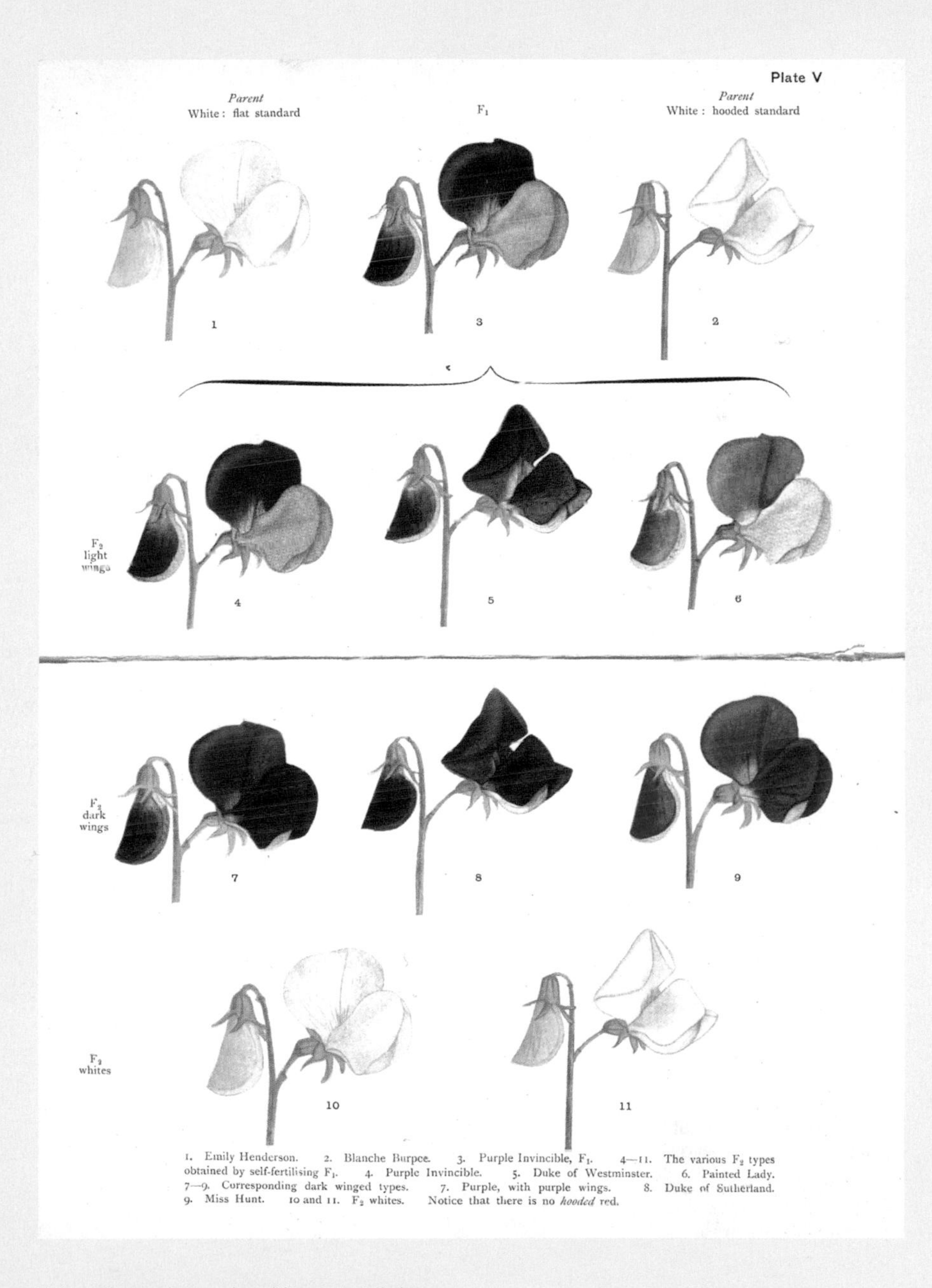


Various differences in generations of the sweet pea (*Lathyrus odoratus*). Plate V from *Mendel's Principles of Heredity* (1909) by William Bateson.

PANSPERMIA **p.156** EVOLUTION BY NATURAL SELECTION **p.161** DNA PROFILING **p.205**
CRISPR GENE EDITING TOOLS **p.206**

# The Nonexistence of Ether

1887

**ALBERT MICHELSON AND EDWARD MORLEY:** *ON THE RELATIVE MOTION OF THE EARTH AND THE LUMINIFEROUS ETHER* • CLEVELAND, OHIO, USA

**Key publications by Albert Michelson**
*The Rigidity of the Earth* 1919
*The Ranges and Phase-displacements of the Earth and Ocean Tides* 1930

**Key publication by Edward Morley**
*On the Densities of Hydrogen and Oxygen and on the Ratio of Their Atomic Weights* 1895

**ALBERT MICHELSON**
Polish American Michelson taught at the US Naval Academy in Annapolis, Maryland, and it was there that he and Morley carried out their experiments. Michelson was the first US Nobel laureate, winning the 1907 prize in Physics.

In the 1860s James Clerk Maxwell (1831–1879) proposed that light was a wave oscillating in an electromagnetic field. It was assumed that, like sound and water waves, light waves needed a material medium. However, this was complicated by the way that light passes through a vacuum (unlike other waves). Physicists asserted that this meant the whole of space was permeated with a subtle and obscure material they termed 'luminiferous ether' – a name that harks back to the days of Aristotle's explanation of the Universe. To detect it, Albert Michelson (1852–1931) and Edward Morley (1838–1923) devised an experiment to measure the 'ether wind'. Their hypothesis was that Earth dragged the ether along with it as the planet ploughed through space, and so the speed of light of beams travelling in different directions should vary slightly depending on their position relative to this drag. The pair tried to show this by splitting a beam of light in two and reflecting each beam through different angles before recombining them. Any differences in speed would make the recombined beam flicker. The experiment revealed no such result. The speed of light stayed firmly constant, a fact that Albert Einstein used in 1905 to disprove the existence of ether.

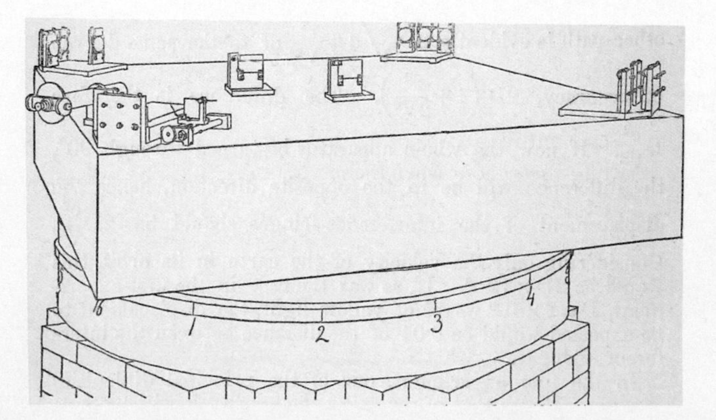


Diagram of an interferometer and four mirrors mounted on a stone floating in mercury. By passing electromagnetic waves through this device, Einstein demonstrated that that form of energy did not need a material medium through which to travel, and he thus disproved the existence of luminiferous ether.

THE NEW PHYSICS **p.27** THE SIZE OF THE UNIVERSE **p.28**

# The Function of Chromosomes

1890s

**THEODOR BOVERI:** *ERGEBNISSE ÜBER DIE KONSTITUTION DER CHROMATISCHEN SUBSTANZ DES ZELLKERNS* • MUNICH, GERMANY

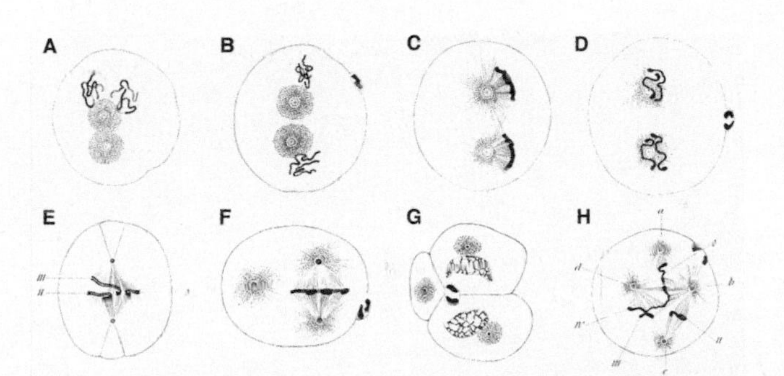


Boveri's drawings of abnormal cleavage divisions in ascaris, roundworm parasites that may live in the human intestine.

**Key publication by Theodor Boveri**
*Ergebnisse über die Konstitution der chromatischen Substanz des Zellkerns* 1904

**THEODOR BOVERI**
Boveri qualified as a doctor in Munich and began work on cell biology at the city's zoological institute. As well as his work on chromosomes, he gave the first account of meiosis, the process by which a body cell divides into sex cells, the sperm and eggs. Additionally, he made the first descriptions of cancerous cells.

Our understanding of heredity involves two strands of investigation. The first was begun by Gregor Mendel, who studied how inherited characteristics moved from generation to generation. The second was the study of how a cell divides, which is the basis of all reproduction. German biologist Walther Flemming (1843–1905) extended the power of microscopy to observe that thread-like objects appeared in the nucleus as the cells prepared to divide. These were named chromosomes. Flemming's hunch was that the chromosomes divided in two as the cell split, with equal numbers of them moving to each new cell. In the late 1890s, Theodor Boveri (1862–1915) was able to track chromosomes moving between the embryo cells of roundworms. He found that each worm's sperm and egg cells only have two chromosomes each, which combine into a set of four when the sperm fertilizes the eggs. Next Boveri showed, this time in sea urchins, that without chromosomes, embryos did not develop at all. Here was the first proof that the information passed from parents to offspring was held in the chromosomes.

GENETICS p.31 GENETIC MODIFICATION p.38

# Discovery of Electromagnetic Waves

**HEINRICH HERTZ:** *ELECTRIC WAVES: BEING RESEARCHES ON THE PROPAGATION OF ELECTRIC ACTION WITH FINITE VELOCITY THROUGH SPACE* • KARLSRUHE, GERMANY

**Key publication by Heinrich Hertz**
*Untersuchungen über die Ausbreitung der elektrischen Kraft* 1892

In the 1860s, James Clerk Maxwell formulated a mathematical description of the electromagnetic field that linked the forces of magnetism with those that propelled electric currents. His work revealed that light was an oscillation in the field now known as electromagnetic radiation. Light was already a known phenomenon, and had recently been proven to act as an oscillating wave. The invisible rays of heat or infrared and ultraviolet light had also been discovered, but Maxwell predicted that there were other waves with much less and much more energy.

In 1893, Heinrich Hertz (1857–1894) discovered what are now known as radio waves. His apparatus created the waves by sending a powerful spark across a gap between two brass balls. He had to black out his windows and spent weeks in pitch darkness looking for the faintest spark to form as 'electric waves' flew across the room. As Maxwell had predicted, these Hertzian waves moved at the speed of light. Hertz was dead before he could build on his discovery, but many other researchers, chief among them Guglielmo Marconi (1874–1937), were soon learning to use the waves for wireless communication.

THE NEW PHYSICS **p.27**

**HEINRICH HERTZ**

Hertz is one of the most famous physicists because the unit of frequency, the hertz (Hz), is named after him, aptly so because it is often used to describe the characteristics of the radio waves he discovered. Hertz's career was short. He died at the age of 36 from a rare blood disorder, but even at a young age his brilliance was evident and he had already been appointed Director of the Physics Institute in Bonn.

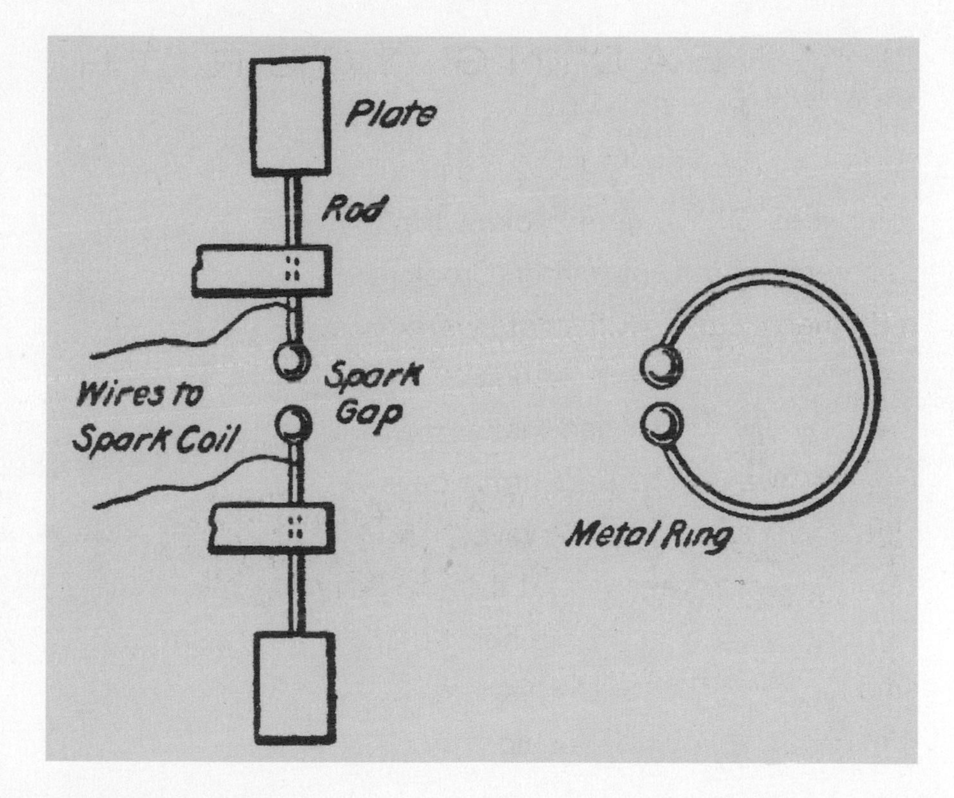


Hertz's 1887 apparatus for generating and detecting radio waves: a spark transmitter (left) consisting of a dipole antenna with a spark gap powered by high voltage pulses from a Ruhmkorff coil, and a receiver (right) consisting of a loop antenna.

FOUR FUNDAMENTAL FORCES **p.165** X-RAY IMAGING **p.194**

# Discovery of Radioactivity

1896

**HENRI BECQUEREL:** *ÉMISSION DE RADIATIONS NOUVELLES PAR L'URANIUM MÉTALLIQUE* • PARIS, FRANCE

**Key publication by Henri Becquerel**
*On Radioactivity, a New Property of Matter* 1903

French physicist Henri Becquerel (1852–1908) had an interest in phosphorescent minerals that give out a mysterious glow. Following the discovery of X-rays in 1895, Becquerel got to wondering whether similar invisible rays were being released from these minerals. He emulated the discovery of X-rays as much as possible. X-rays passed through materials that blocked visible light, so Becquerel sheathed a photographic plate in black paper. He then placed samples of minerals and artificial compounds that also phosphoresced on the paper. If any samples were releasing X-rays, then the hidden plate would still become fogged. In 1896 his endeavour yielded its first result from uranyl, an oxide of uranium. Becquerel then switched his attention to other uranium compounds, even ones that did not glow in the dark. This set of experiments showed that the emissions were firmly linked to uranium, not to phosphorescence. (Today the properties of phosphorescence and fluorescence are understood as light being absorbed by atoms and then remitted.) Physicists from across the world began to study 'Becquerel rays', as the phenomenon was dubbed. Soon thorium was found to also emit them, and Marie Curie (1867–1934) began finding undiscovered elements with the same property. It was Curie who described the phenomenon that became known as radioactivity – the active release of radiation.

**HENRI BECQUEREL**
The Becquerels were a nineteenth-century scientific dynasty in France. Henri's grandfather invented an early form of battery, his father discovered the photovoltaic effect where light shining on a surface makes electricity, and Henri also found a link between the magnetic and optical properties of crystals. All four held the physics chair at the Muséum national d'Histoire naturelle in Paris. Henri Becquerel has the unit of radioactivity named for him. He shared the 1903 Nobel Prize for his discovery with Marie and Pierre Curie (1859–1906).

THE NEW PHYSICS **p.27** SCIENCE AND THE PUBLIC GOOD **p.29**

Marie Curie and her daughter, Irène, in 1925. In 1903, the former became the first woman to win the Nobel Prize and in 1911 the first person to win the award in two different fields (the first was for Physics, the second Chemistry). In 1935 Irène shared the Chemistry award with her husband, Frédéric Joliot-Curie.

FOUR FUNDAMENTAL FORCES **p.165** X-RAY IMAGING **p.194** GEIGER-MÜLLER TUBE **p.191**
RADIOCARBON DATING **p.197** SCHRÖDINGER'S CAT AND OTHER THOUGHT EXPERIMENTS **p.210**

# Discovery of the Electron

1897

**J.J. THOMSON:** *ON THE MASSES OF THE IONS IN GASES AT LOW PRESSURES*
CAMBRIDGE, UK

**Key publications by J.J. Thomson**
*Electricity and Matter* 1903
*James Clerk Maxwell* 1931

By the end of the nineteenth century, what we now call a cathode-ray tube proved to be a very significant apparatus. The basic design was a glass tube fitted with an anode and a cathode (a positive and a negative electrode) inside. Then the air inside it was voided to make a near-vacuum. When electrified, the device produced a cathode ray, a mysterious beam that came out of the cathode and glowed in the dark. Already a souped-up tube had revealed the existence of X-rays, and in 1897 J.J. Thomson (1856–1940) returned with yet another improved version to investigate the rays. His high-powered kit revealed that the beam did swing towards a positive electric charge, thus showing it to be negatively charged. Light has no charge, so Thomson proposed that this was a beam of particles instead. He then calculated the mass of each particle, and found that it was 1,800 times lighter than a hydrogen atom. That meant these were subatomic particles smaller than any atom. This was something thought to be impossible, but the particles, which Thomson named 'electrons', proved to be the missing link in understanding electric current. Soon there was a growing family of other subatomic particles that revealed the structure and properties of the atom.

**JOSEPH JOHN 'J.J.' THOMSON**

**Joseph John 'J.J.' Thomson was awarded the 1906 Nobel Prize for his discovery. As a young mathematician and physicist, he had impressed enough to be elected Cavendish Professor at Cambridge before the age of 40, and he did not rest on his laurels after his subatomic breakthrough. In 1912 he was a co-discoverer of isotopes (chemically identical elements with different masses), and he combined all this knowledge to build the first mass spectrometer, a device for analyzing atoms and molecules by their size.**

ELECTRICITY **p.24** THE NEW PHYSICS **p.27**

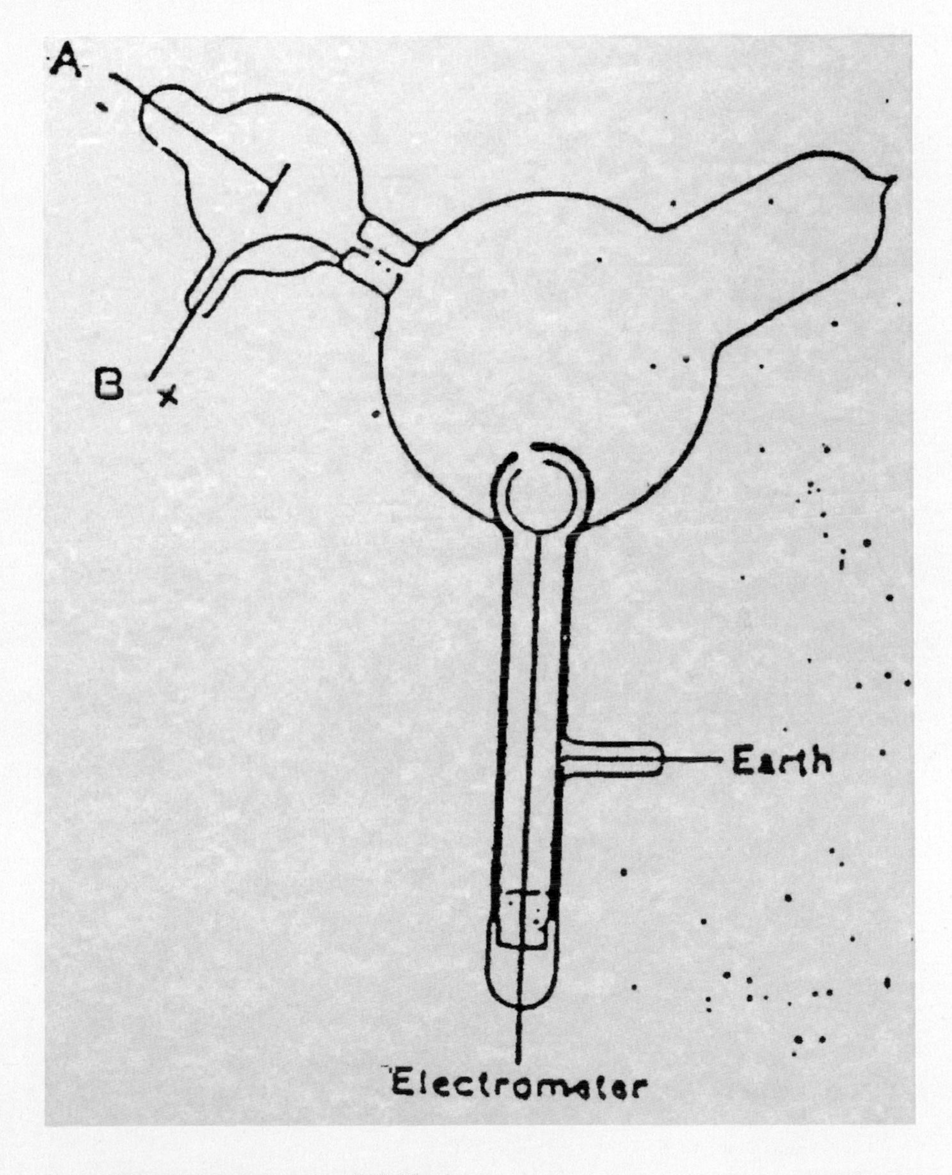


Diagram of Thomson's cathode-ray tube in an 1897 edition of *Philosophical Magazine*. Rays from the cathode (A) are passed into the bulb through a slit in a metal plug in the neck of the tube; the plug is connected to the anode (B).

QUANTUM PHYSICS **p.167** VALENCE BOND THEORY **p.168** THE STANDARD MODEL **p.174**
CATHODE-RAY TUBE **p.193** MASS SPECTROMETRY **p.202**

# Learned Responses

1898 – 1930

**IVAN PAVLOV:** *THE EXPERIMENTAL PSYCHOLOGY AND PSYCHOPATHOLOGY OF ANIMALS* • ST PETERSBURG, RUSSIA

Ivan Pavlov (centre, with grey beard) with assistants and students at the Imperial Military Academy of Medicine, St Petersburg, prior to a demonstration of his experiment on a dog (c.1912–1914).

NATURAL HISTORY AND BIOLOGY **p.22** NEUROSCIENCE AND PSYCHOLOGY **p.34**

Ivan Pavlov's (1849–1936) name has become synonymous with the way animals learn. A Pavlovian response is one where an animal's behaviour is changed by a neural stimulus. Famously, Pavlov was able to make dogs salivate, as they do when preparing to eat food, after hearing just a bell (and not by providing food). This field of research was arrived at by accident. Pavlov was already a world expert on digestion, and was investigating the nerve link between brain and stomach, especially how it caused the production of stomach juices. He set up an experiment to collect the saliva dripping from the jaws of a team of captive dogs. He noticed that the dogs salivated more when their food was brought in by his assistants – nothing unusual about that – but crucially they also salivated when the assistant came in empty-handed on another errand. The dogs had learned the link between food and the assistants, and were responding to the latter in the same way as they would to the former. Pavlov recreated this effect by ringing a bell at feeding time, and soon the dogs salivated at the sound of the bell. Pavlov reasoned that salivating at the sight of food was an instinctive or reflex response, and named the learned response a 'conditioned reflex'. As a result, this simple form of learning by associating stimuli with rewards or punishments is now known as conditioning.

**IVAN PAVLOV**

**The son of a priest in a provincial city south of Moscow, Pavlov entered a theological seminary after school but soon switched to science. In 1890 he became director of physiology at the Institute of Experimental Medicine in St Petersburg, and spent the rest of his life working there. In 1904 he was awarded the Nobel Prize in Medicine before the work on conditioning for which he is now famous was widely known.**

**Key publications by Ivan Pavlov**

***The Work of the Digestive Glands*** 1902

***Conditioned Reflexes: An Investigation of the Physiological Activity of the Cerebral Cortex*** 1927

***Lectures on Conditioned Reflexes: Twenty-five Years of Objective Study of the High Nervous Activity (Behavior) of Animals*** 1928

MACHINE LEARNING **p.213**

# Sex Chromosomes

1905

**NETTIE STEVENS:** *STUDIES IN SPERMATOGENESIS PART II, A COMPARATIVE STUDY OF THE HETEROCHROMOSOMES IN CERTAIN SPECIES OF COLEOPTERA, HEMIPTERA, AND LEPIDOPTERA, WITH ESPECIAL REFERENCE TO SEX DETERMINATION* • WASHINGTON, D.C., USA

**Key publication by Thomas Hunt Morgan**
***Sex Limited Inheritance in Drosophila*** 1910

Chromosomes come in pairs, one inherited from the mother and the other from the father. Thus every body cell carries a double set of chromosomes. When the body makes sex cells (sperm or eggs), the number of chromosomes is halved, with each sperm or egg having one chromosome from each pair. During conception, the half-sets from each sex cell combine to make a brand-new full set of chromosomes.

In humans, the chromosome number is 46, or 23 pairs, and 22 of the pairs are made up of matching chromosomes of the same length. The twenty-third pair does not always match like for like. Females have a pair of long chromosomes, denoted as X, while a male cell contains one X and a much smaller Y chromosome. That means that eggs (female sex cells) always carry one X chromosome, while a sperm (the male sex cell) has either an X or a Y chromosome. On fertilization, the sperm adds its X or Y chromosome to the egg to determine the sex of the baby. This system was discovered in 1905 by Nettie Stevens (1861–1912), an American cell biologist. She saw that the sperm of mealworms – beetle larvae – had either a large or a small chromosome. The XY system occurs widely in mammals, insects and many reptiles. Birds have a ZW system, which works in reverse, with females having the mixed pair of chromosomes.

**THOMAS HUNT MORGAN**

**Five years after the discovery of sex chromosomes, Thomas Hunt Morgan (1866–1945) showed that particular inherited traits were sex-linked. This means genes that are inherited on the X chromosome are not represented on the smaller Y. As a result, these features show up more often in males and are rarer in females, who generally inherit a dominant alternative version of their second X. Sex-linked features in humans include haemophilia and colour blindness.**

CELL THEORY **p.25** GENETICS **p.31** GENETIC MODIFICATION **p.38**

American geneticist Nettie Stevens at work in 1909 at the Zoological Station in Naples, Italy. She discovered sex chromosomes in 1905.

PANSPERMIA **p.156** DNA PROFILING **p.205**

# Measuring Charge

1909

**ROBERT MILLIKAN:** *ON THE ELEMENTARY ELECTRICAL CHARGE AND THE AVOGADRO CONSTANT* • CHICAGO, ILLINOIS, USA

In 1909, two American physicists devised an experiment to show that, contrary to the prevailing views of the day, electricity was not a flow of charged particles but a wave.

To do that Robert Millikan (1868–1953), assisted by Harvey Fletcher (1884–1981), set about measuring electrical charge. They predicted that the charge would not be a fixed value but would oscillate between two extremes. The method was the now famous oil-drop experiment, where fine droplets of oil were sprayed into a strong electric field between two horizontal plates. The scientists watched the droplets through a microscope as they fell under gravity to the lower plate. Some of the droplets picked up a charge due to friction as they were sprayed, and so were pushed back into the air by the electric field. By measuring how quickly a drop of a certain size and weight moved, the researchers were able to compare the strength of the electromagnetic force against the force of gravity, and then figure out how much charge the droplet held. Many iterations later they found that every result was a multiple of $1.5 \times 10^{-19}$. So Millikan's hypothesis was false: charge does not fluctuate. More significantly, electric charge is indeed carried by particles, such as the electron, and always exists in multiples of a fixed amount (a quantum), a fact that fed into the growing field of quantum physics.

Robert Millikan at work c.1930.

THE NEW PHYSICS p.27 THE SIZE OF THE UNIVERSE p.28

**Key publications by Robert Millikan**

*The Electron: Its Isolation and Measurement and the Determination of Some of its Properties* 1917

*The Autobiography of Robert Millikan* 1950

**ROBERT MILLIKAN**

Millikan grew up in the Midwest of the United States. He was the first person to gain a PhD from the newly established physics department at Columbia University in New York. Millikan began his professional academic career in Chicago, and performed the oil-drop experiment there. In later years he investigated the photoelectric effect, where light shining on substances creates electric current.

ATOMIC THEORY **p.159** QUANTUM PHYSICS **p.167** VALENCE BOND THEORY **p.168** BUBBLE CHAMBERS **p.198**

# Hertzsprung-Russell Diagram

1911

**EJNAR HERTZSPRUNG:** *ZUR STRAHLUNG DER STERNE* • COPENHAGEN, DENMARK

In the early years of the twentieth century, comprehensive surveys of the sky were being published. These were not only upgrades of the star maps that had been made since ancient times, showing the relative locations of stars and other celestial objects. They also included accurate records of stars' brightness, or magnitude, and the colour or colours of light they emitted. Additionally, new techniques in measuring interstellar distances meant that astronomers were able to assign absolute sizes to stars based on how bright they appeared from Earth.

In 1913 two astronomers working independently, the Danish Ejnar Hertzsprung (1873–1967) and the American Henry Norris Russell (1877–1957), sought to make sense of all this information. They knew that the colour of stars is indicative of their temperature – blue and white stars are hotter than yellow and orange; red stars are the coolest – and they found that the Universe's stars were not randomly scattered across this chart. Instead, most stars formed the 'main sequence', which was a zone running diagonally across the diagram from hot and large to cold and small. Our Sun sits in a very average middle position. Elsewhere, clusters of giant and supergiant stars tended to be cooler, while the smallest stars, white dwarfs, were all very hot indeed. The Hertzsprung-Russell Diagram opens a window into the lifecycle of stars. The great majority of stars spend billions of years in the main sequence. Then, as they run out of fuel for the fusion that produces their heat and light, the stars will swell into bigger but cooler red giants. Finally, the giant star disintegrates, leaving behind a tiny, hot core known as a white dwarf star.

**EJNAR HERTZSPRUNG**

Born in Copenhagen but working for most of his career in Germany and the Netherlands, Hertzsprung is a somewhat unsung astronomer despite being at the heart of the momentous advances that science made in his day. He helped to perfect the process of measuring distances between stars that allowed Hubble to discover universal expansion. Hertzsprung was the son-in-law of Jacobus Kapteyn (1851–1922), who showed that the Milky Way is rotating. Hertzsprung also taught Gerard Kuiper (1905–1973), who was a founding figure of planetary science and for whom the Kuiper belt of ice objects beyond Neptune is named.

ANCIENT ASTRONOMERS **p.12** THE SIZE OF THE UNIVERSE **p.28** THE UNIVERSE IS MISSING **p.37**

**Key publication by Ejnar Hertzsprung**
*Publikationen des Astrophysikalischen Observatoriums zu Potsdam* 1911

**Key publications by Henry Norris Russell**
*New Regularities in the Spectra of the Alkaline Earths* 1925
*On the Composition of the Sun's Atmosphere* 1929

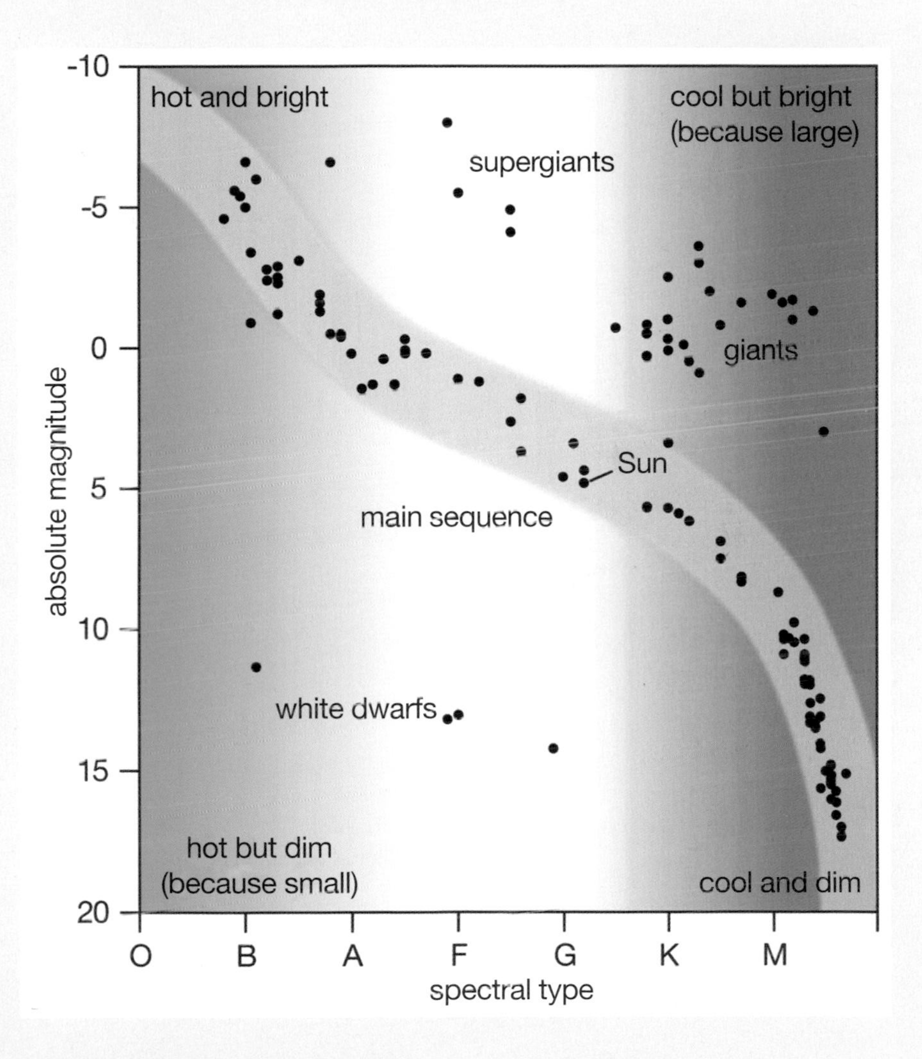


The Hertzsprung-Russell Diagram. Spectral type (a measure of a star's temperature) is plotted on the horizontal axis, and absolute magnitude (the intrinsic brightness of a star) is plotted on the vertical axis.

BIG BANG **p.169** STELLAR NUCLEOSYNTHESIS **p.170** TELESCOPES **p.189**

# Cosmic Rays

1911

**VICTOR HESS:** *ON THE OBSERVATIONS OF THE PENETRATING RADIATION DURING SEVEN BALLOON FLIGHTS* • PIESKOW, GERMANY

By 1910 physicists were experimenting with ever-more sensitive electroscopes (devices for measuring electric charge). They found that the air grew more highly charged with altitude, with one notable experiment being carried out at the top of the Eiffel Tower in Paris. In 1911, Victor Hess (1883–1964) began a series of high-altitude balloon flights to survey the electrical properties of the atmosphere. He charged up electroscopes before take-off, and found that this charge dissipated more quickly the higher he flew.

The conclusion Hess drew was that high-altitude air was a better electrical conductor than air at sea level, with a greater number of charged particles able to draw away charge from the electroscope. Hess's suggestion was that the molecules in the air were becoming charged (ionized) by collisions with high-speed particles smashing into the atmosphere from space. Rather confusingly, these streams of particles have become known as cosmic rays. The particles range in size and charge and originate in the Sun or from outside the Solar System, and they can arrive at Earth travelling at speeds close to the speed of light.

Hess and others recognized that observing collisions with cosmic rays was a good way of seeing how atoms were constructed and behaved. The first exotic subatomic particle, the muon, was initially detected in these high-altitude collisions, and that inspired the construction of particle accelerators to mimic these effects in a more controlled manner.

**VICTOR HESS**
Born in central Austria, Hess performed his flights of discovery while serving at the Viennese Academy of Science. In 1921, he moved to the United States and worked as a mining consultant before returning to Europe to begin a career as a physics professor. The arrival of the Nazis in Austria led him to move back to the United States. He spent the rest of his career at Fordham University in New York.

**Key publication by Victor Hess**
*The Electrical Conductivity of the Atmosphere and Its Causes* 1928

ELECTRICITY **p.24** STRING THEORY **p.39**

Victor Hess returning from one of his balloon flights
in August 1912.

BUBBLE CHAMBERS **p.198** PARTICLE ACCELERATORS **p.199** ATLAS (CERN) **p.200**

# Atomic Nucleus

1917

**ERNEST RUTHERFORD:** *THE SCATTERING OF α AND β PARTICLES BY MATTER AND THE STRUCTURE OF THE ATOM* • MANCHESTER, UK

**Key publications by Ernest Rutherford**

*Radio-activity* 1904

*Radioactive Transformations* 1906

*Radioactive Substances and their Radiations* 1913

*The Electrical Structure of Matter* 1926

*The Artificial Transmutation of the Elements* 1933

THE NEW PHYSICS **p.27** STRING THEORY **p.39**

The discovery in 1897 of the electron, the first subatomic particle, raised the question of what else the atom was made of. Electrons are negatively charged, while overall an atom has no charge, so the initial idea was the 'plum-pudding model', in which electrons were spread through a positively charged matrix, like the plums in a pudding. By 1909, research into radioactivity had found that atoms could give out positively charged 'alpha particles'. Ernest Rutherford (1871–1937) saw this as a contradiction to the plum-pudding model and set up an experiment with his students Hans Geiger (1882–1945) and Ernest Marsden (1889–1970). They fired alpha particles at a thin gold foil and set up photographic plates beyond to pick up each particle as a fogged dot. The particles mostly sailed right through the foil, with a few deflected to left and right. The team relocated the detectors to behind the alpha particle source. This revealed that a tiny number of the particles were bouncing right back. The positive particles were being repulsed by the positive component of the atom. Rutherford analyzed the path of all particles to show that all the atom's positive charge was concentrated in a tiny core, which he named its nucleus. In 1917, Rutherford discovered that the positive charge was carried by subatomic particles, which he named protons, that are packed into the nucleus.

**ERNEST RUTHERFORD**

**A New Zealander by birth, Rutherford began his research career in Montreal, Canada, where he differentiated the radiation detected from radioactive elements into two forms: alpha radiation, which was positive and made of heavy particles (now known to be helium nuclei), and beta radiation, which has a negative charge and was lighter in weight (and is now identified as consisting mainly of electrons). He then moved to Manchester, England, where he performed the nucleus experiment, and later became the head of the Cavendish Laboratory at Cambridge. As well as finding the nucleus and the proton, he also oversaw the discovery of the neutron. Rutherford is buried near Isaac Newton in Westminster Abbey, London.**

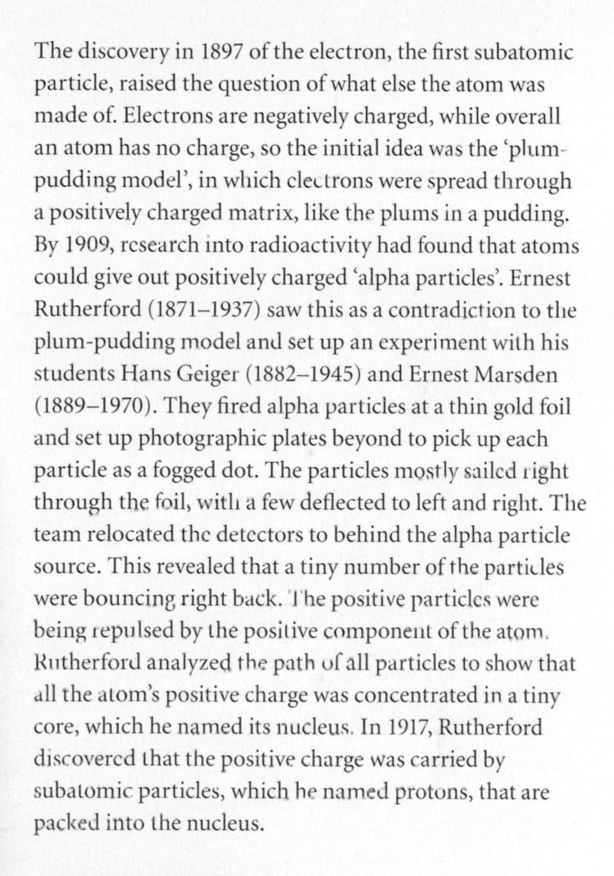

Hans Geiger (left) and Ernest Rutherford at work in 1937 in a laboratory at McGill University, Montreal.

ATOMIC THEORY **p.159** THE STANDARD MODEL **p.174** ATLAS (CERN) **p.200**

# Wave-Particle Duality

1927

**GEORGE THOMSON:** *DIFFRACTION OF CATHODE RAYS BY A THIN FILM*
ABERDEEN, UK

**Key publications by George Thomson**

*The free path of slow protons in helium* 1926

*Electron optics* 1932

*The growth of crystals* 1948

Few children of Nobel laureates have managed to equal their parent's achievement, but George Thomson (1892–1975), son of J.J., did just that in 1937. His award came for research in the same field as his father. J.J. discovered electrons; George showed that they are waves. He did this by performing an alternative version of Thomas Young's 1804 double-slit experiment, which showed that light is a wave. To this end, Thomson fired a beam of electrons at a thin layer of metal foil. Particles would have passed through the foil and been picked up as a random spread on the detector beyond. However, Thomson found the stream of electrons created an interference pattern on the detector as if they were behaving like waves. This seemingly bizarre result had been predicted in the 1920s by the founders of quantum mechanics. They found a way to describe the behaviours of electrons and other subatomic objects by treating them as wave-like entities with properties that fluctuate, or oscillate, from one state to another. The fact that electrons can behave like particles and waves at the same time is a phenomenon now known as wave-particle duality.

**GEORGE THOMSON**

To avoid confusion with several other George Thomsons in science and other fields, this particle physicist is generally referred to as George Paget Thomson. In World War I he worked on aerodynamics for the Royal Flying Corps, the precursor to the RAF. After the war he moved to the University of Aberdeen, where he carried out his work on electrons. In World War II Thomson led British research into atomic weapons. This work was passed to the United States and was a key foundation of the Manhattan Project, which eventually produced nuclear weapons in 1945.

THE NEW PHYSICS **p.27** SCIENCE AND THE PUBLIC GOOD **p.29** STRING THEORY **p.39** 

This photograph by Thomson shows an electron at the centre of the image undergoing diffraction.

QUANTUM PHYSICS p.167 PARTICLE ACCELERATORS p.199

# Antibiotics

1928

**ALEXANDER FLEMING:** *ON THE ANTIBACTERIAL ACTION OF CULTURES OF A PENICILLIUM, WITH SPECIAL REFERENCE TO THEIR USE IN THE ISOLATION OF B. INFLUENZAE* • LONDON, UK

**Key publications by Alexander Fleming**
*The Physiological and Antiseptic Action of Flavine* 1917
*On a remarkable bacteriolytic element found in tissues and secretions* 1922

**ALEXANDER FLEMING**
Fleming qualified as a doctor then served in World War I. For his part in the development of penicillin, he shared the 1945 Nobel Prize with Florey and Ernst Chain (1906–1979).

By the 1880s the link between bacteria and infections was firmly established and the idea of the 'magic bullet' was born. While several chemicals were known to be antiseptic, meaning they killed germs, they were also damaging to healthy tissues. A magic bullet would kill the germs without hurting the body. Various chemical candidates had been tried, but the first truly effective form of this medicine, now known as an antibiotic, was discovered by accident in 1928 by Scottish bacteriologist Alexander Fleming (1881–1955). Fleming returned from a summer break to find that, although blue-green mould had spread across his bacteria samples, the region around the fungus was clear of bacteria. He carried out experiments to confirm that the fungus was producing an antibiotic chemical, which he extracted and named penicillin. He found that penicillin was not toxic to humans, and saw immediately that it would be useful as a magic bullet for fighting infections. He even demonstrated this by treating a colleague's eye infection with fungal dust. However, few other scientists were convinced and biomedical engineering at the time was unable to mass-produce the active ingredient. In 1940 Howard Florey (1898–1968) developed a way of manufacturing penicillin as a broth, and carried out a clinical trial to prove that the antibiotic was safe. Penicillin and other antibiotics are estimated to have since saved 200 million lives.

THE BIRTH OF MEDICINE **p.14** PUBLIC HEALTH **p.26**

Alexander Fleming, Professor of Bacteriology at London University, in his laboratory at St Mary's Hospital (1943).

CLINICAL TRIALS **p.208**

# The Expanding Universe

1929

**EDWIN HUBBLE:** *A RELATION BETWEEN DISTANCE AND RADIAL VELOCITY AMONG EXTRA-GALACTIC NEBULAE* • MOUNT WILSON, CALIFORNIA, USA

**EDWIN HUBBLE**

Despite a general interest in maths and science, Hubble acquiesced in his father's desire that he study law. He excelled and won a Rhodes Scholarship to Oxford. After his father's death in 1913, Hubble became a teacher before completing a PhD in astronomy. He served a year in the US Army in World War I before joining the Mount Wilson Observatory, where he worked until his death.

Edwin Hubble (1889–1953) is famous for two immense discoveries. The first, in 1924, showed that our galaxy, the Milky Way, did not comprise the whole Universe but was just one of many other galaxies (now estimated at two trillion) separated by the vast emptiness of intergalactic space. Hubble discovered this by making surveys of the sky from the Mount Wilson Observatory in California, which had the largest telescope in the world at the time. His discovery relied on the work of Henrietta Swan Leavitt (1868–1921) who had found how to measure the distance to a type of star that varied in brightness. How quickly it faded and brightened showed how large it was. Once the size was known, the apparent brightness of the star from Earth told observers how far away it was. Hubble used this process to show that some stars were beyond the Milky Way. That meant that many fuzzy nebulous objects were shown to be galaxies separate from ours. It was already known that these bodies were all moving away from us, and by 1929 Hubble's continued survey showed that the ones that were further away were moving faster. In astronomy, more distant objects are also older, and so the distances between us and older objects are increasing faster than the distances between us and younger objects: in other words, the whole of space is steadily expanding.

**Key publications by Edwin Hubble**

*The Color of the Nebulous Stars* 1920

*Extragalactic Nebulae* 1926

*Redshifts in the Spectra of Nebulae* 1934

ANCIENT ASTRONOMERS **p.12** THE SIZE OF THE UNIVERSE **p.28** THE UNIVERSE IS MISSING **p.37**

Edwin Hubble using the 254-centimetre (100-inch) Hooker telescope at Mount Wilson Observatory, Pasadena, California.

BIG BANG **p.169** DARK MATTER **p.175** COSMIC INFLATION **p.176**

# Recombination

1931

**BARBARA MCCLINTOCK:** *THE ORIGIN AND BEHAVIOR OF MUTABLE LOCI IN MAIZE* • ITHACA, NEW YORK, USA

**Key publication by Barbara McClintock**
*The Barbara McClintock Papers*
1927–1991

In 1931 American geneticist Barbara McClintock (1902–1992) was studying the chromosomes of maize and found that, in some strains of the plant, one of the chromosomes had a knob on the end while its matching pair did not. (Chromosomes are always in homologous pairs. Both parents provide one of each pair, and both chromosomes contain different versions of a particular set of genes.) McClintock found that the knob was associated with the colour and starchiness of the maize seed, and used the knob to track chromosomes through the process of meiosis. (Meiosis is an unusual cell division that splits up the homologous pairs, creating sex cells with half the regular number of chromosomes. Sex cells for two parents combine their chromosomes during fertilization to create a new full set.)

Amazingly, McClintock found that the knob structure moved between paired chromosomes during meiosis, the first evidence of a process now known as recombination. Recombination sees paired chromosomes line up next to each other during meiosis and cross their DNA strands, so sections can be exchanged. The result is that chromosomes inherited from the parents are jumbled up by the offspring, creating a uniquely different set of chromosomes with each passing generation. For many years McClintock's work was ignored, mainly because it was way ahead of its time. Nevertheless, in 1983 – 52 years later – Barbara McClintock received the Nobel Prize for her discovery of 'jumping genes'.

**BARBARA MCCLINTOCK**
McClintock began her career in 1927 as a botanist, and later entered the emerging field of cytogenetics – the study of the cellular mechanisms behind genetics. Her breakthrough in the 1930s led to another 20 years of research. However, in 1953 she gave up publishing results due to the indifference of her colleagues. It was not until the 1970s that the significance of her discoveries was finally recognized.

CELL THEORY **p.25** GENETICS **p.31** GENETIC MODIFICATION **p.38**

Barbara McClintock at work in her laboratory in the Department of Genetics, Carnegie Institution, Cold Spring Harbor, New York.

PANSPERMIA **p.156** DNA PROFILING **p.205** CRISPR GENE EDITING TOOLS **p.206**

# Nuclear Fission

1934

**ENRICO FERMI:** *NUOVI RADIOELEMENTI PRODOTTI CON BOMBARDAMENTO DI NEUTRONI* • ROME, ITALY

Inspired by the way that Ernest Rutherford had used alpha particles to unlock secrets of the nucleus, Enrico Fermi (1901–1954) had the idea to fire neutrons, which had only recently been discovered, at atoms to see what happened. Neutrons have no charge, unlike alpha particles, so they are not deflected by charged electrons and protons in the target atoms. In 1934, Fermi's team announced that by bombarding uranium they had created a new element. Fermi named it hesperium, but it is now known as plutonium. German scientist Otto Hahn (1879–1968) repeated Fermi's experiments and found that he had made barium. Hahn's colleague Lise Meitner (1878–1968) showed that the barium was created by the extra neutron making the uranium nucleus unstable, so it split into two smaller atoms. This was nuclear fission. Hungarian Leo Szilard (1898–1964) found that if one atom released free neutrons during fission, then they could cause more fissions in a chain reaction. For its size, a fissioning atom releases a vast amount of energy, and Szilard realized that an uncontrolled fission reaction could be used as a bomb. In 1942, Fermi set up Pile 1, the world's first nuclear reactor, at Chicago University. Blocks of graphite were used to absorb and focus neutrons and so create a slow chain reaction.

**ENRICO FERMI**

**Born in Rome but spending much of his childhood in the Italian countryside, Fermi showed his exceptional intellect early on. By the age of 20 he had published his first scientific discovery, and aged 24 he became a professor of physics in Rome. It was there that he became a world-leading atomic physicist. Forced out of Europe by the rise of fascism, Fermi moved to the United States in 1938. After setting up Pile 1, he declined to be involved in the Manhattan Project's search for nuclear weapons. Fermi died young from cancer due to repeated exposure to radioactivity.**

NATURAL HISTORY AND BIOLOGY **p.22** THE NEW PHYSICS **p.27** SCIENCE AND THE PUBLIC GOOD **p.29**

The world's first nuclear reactor, Pile 1, erected in 1942 in the West Stands section of Stagg Field at the University of Chicago.

**Key publications by Enrico Fermi**

*Introduzione alla fisica atomica* 1928

*Molecole e cristalli* 1934

*Elementary particles* 1951

FOUR FUNDAMENTAL FORCES p.165 GEIGER-MÜLLER TUBE p.191

# Turing Machine

1936

**ALAN TURING:** *ON COMPUTABLE NUMBERS, WITH AN APPLICATION TO THE ENTSCHEIDUNGSPROBLEM* • CAMBRIDGE, UK

**Key publication by Alan Turing**
***Computing Machinery and Intelligence*** 1950

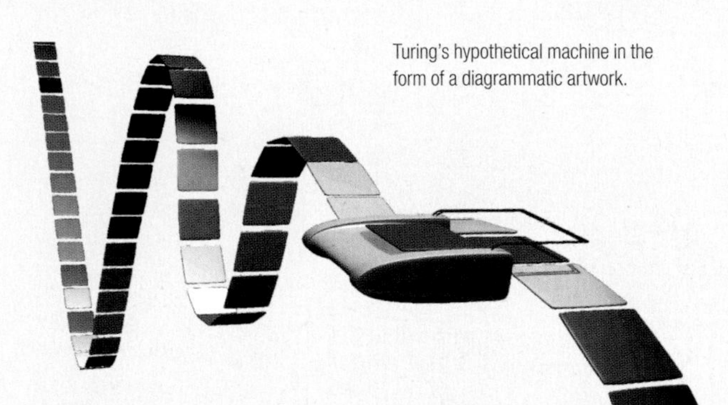

Turing's hypothetical machine in the form of a diagrammatic artwork.

The basic functions of the modern digital computer were by-products of a thought experiment performed in 1936 by English mathematician Alan Turing (1912–1954). Turing wanted to tackle the Decision Problem. This was a mathematical mystery which wanted to know if a logical process, or algorithm, would arrive at an answer or loop forever without reaching an endpoint. The Decision Problem asked if there was a way of telling the first type of algorithm from the second without just running the process to find out. In order to investigate, Turing imagined a 'virtual machine' that had a strip of tape with symbols or data on it. The tape never ran out and could move left or right beneath a head that could read the data and rewrite it. The behaviour of the head and the tape was controlled by a list of rules. This list was the algorithm. It responded to the data on the tape and then gave a corresponding command, such as 'move the tape', 'rewrite data', etc. Depending on the algorithm used, the Turing machine would either halt on one answer or continue forever. Turing's imaginary device showed that the answer to the Decision Problem was no, there is no way to discern a halting algorithm from a looping one. The Turing machine became the model for the way today's computer hardware is structured and handled by software algorithms (which don't loop!).

**ALAN TURING**

**Few people had as big a personal impact on modern life as Turing. His work on computing was compounded by the famous Turing Test (or Imitation Game) for artificial intelligence. Additionally, he was an instrumental figure in using primitive computing power to decode enemy codes in World War II, shortening the conflict and saving an estimated 14 million lives. Turing was gay, which was a crime in Britain at the time. A conviction in 1952 led to his removal from high-level computing research, which contributed to Turing taking his own life by cyanide poisoning.**

ELECTRONICS AND COMPUTATION **p.30** THE INTERNET **p.36**

# Citric Acid Cycle

1937

**HANS KREBS:** *THE ROLE OF CITRIC ACID IN INTERMEDIATE METABOLISM IN ANIMAL TISSUES* • SHEFFIELD, UK

**Key publications by Hans Kreb**
*Metabolism of ketonic acids in animal tissues* 1937
*Energy Transformations in Living Matter: A Survey* 1957
*Reminiscences and Reflections* 1981

**HANS KREBS**
In his native Germany Krebs tried to investigate respiration but his work was often blocked by his supervisor. In 1933, as a Jew, Krebs was forced to flee the Nazis and eventually settled in England, where he was free to pursue the research that made his name. The cycle is often referred to as the Krebs cycle.

In the late 1700s it was discovered that living things take in oxygen and give it back out as carbon dioxide. This conversion revealed that organisms essentially burn food to release the energy needed to power life. The process is known as respiration. However, outside of the body that simple burning reaction releases energy in a flame.

In 1937, Hans Krebs (1900–1981), a German working in Sheffield, England, completed a series of experiments to show that respiration involved a cyclical metabolic pathway that released energy in a controlled, step-by-step fashion. This discovery showed that nearly all forms of life on Earth – from bacteria to whales – are sustained in this way. The cycle can be fed by sugars, fats or proteins, which are all broken up into a simple substance named acetyl coenzyme A, molecules of which bond to the final product of the pathway, oxaloacetate, to make citrate (a non-acidic version of citric acid). The citrate then cycles through eight steps, gradually removing carbon atoms to make waste carbon dioxide. Each cycle produces two molecules of carbon dioxide. Every successive molecule in the cycle is giving away a little energy. Eventually, a new oxaloacetate is made, and the cycle begins again. Krebs discovered the steps by feeding oxygen to pigeon liver for steadily increasing periods of time and then analyzing the chemicals present. The later components of the cycle became more abundant as the oxygen was supplied for longer.

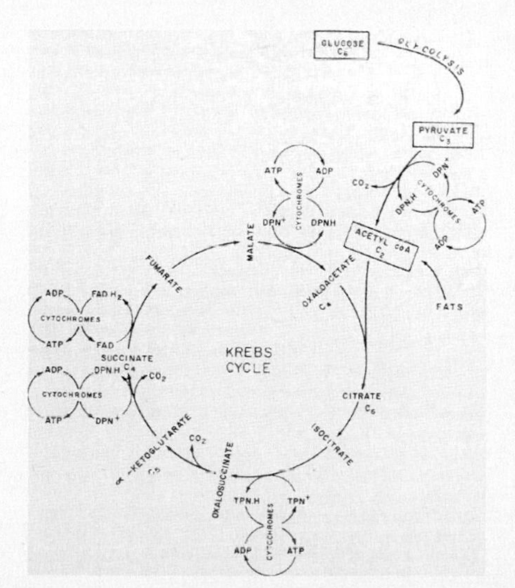


Diagram showing the detailed workings of the Krebs cycle, the discovery of which aided cytology (the study of the structure and function of individual cells).

NATURAL HISTORY AND BIOLOGY **p.22** CELL THEORY **p.25**

# Origin of Life

1953

**STANLEY MILLER:** *A PRODUCTION OF AMINO ACIDS UNDER POSSIBLE PRIMITIVE EARTH CONDITIONS* • CHICAGO, ILLINOIS, USA

All life uses the same set of biochemicals, including sugars, fats and proteins. Today these substances are manufactured by metabolism inside living cells. How did they first arise? The answer to that question is still a mystery, but in 1953 Harold Urey (1893–1981) teamed up with Stanley Miller (1930–2007), both chemists, to test the idea that these materials could be made spontaneously by non-biological processes. This idea was based on the concept of primordial soup, in which life arose from an ocean filled with chemicals. Miller and Urey built an apparatus which they thought would create the conditions of the early Earth. Nicknamed the Lollipop, it comprised a tank half-filled with water, which was heated gently until the liquid evaporated, and the resulting vapour circulated into a gas chamber which also contained methane, hydrogen and ammonia. The gas mixture was periodically electrified to emulate lightning, and then cooled to create a rain-like condensation, which was returned to the water tank. Round and round this material went, and within a day the clear mixture had gone pink. After a week, 10 per cent of the methane had combined into amino acids and other complex organic chemicals. These are the building blocks of proteins. Miller said his experiment produced 11 of the 20 amino acids used in life today.

**HAROLD UREY**

**Urey came to the subject of primordial soup late in his career. He was already a world-leading physical chemist, having won the 1934 Nobel Prize for his discovery of deuterium, a heavy form of hydrogen. (Water made with deuterium is named 'heavy water'.) Urey had a leading role in the Manhattan Project, which developed the atomic bomb, and after World War II he looked for another area of study, settling on cosmochemistry, the study of chemicals found in space. This area led him to wonder what chemicals were present when Earth formed.**

NATURAL HISTORY AND BIOLOGY **p.22** CELL THEORY **p.25** GENETICS **p.31**

Stanley Miller with the apparatus used in the ground-breaking experiment he conducted with Harold Urey.

**Key publication by Stanley Miller**
***The Origins of Life on the Earth*** 1974

PANSPERMIA **p.156** EVOLUTION BY NATURAL SELECTION **p.161** RADIOCARBON DATING **p.197**
STEM CELLS **p.207** CLADISTICS AND TAXONOMY **p.209**

# The Double Helix

1953

**JAMES WATSON AND FRANCIS CRICK:** *MOLECULAR STRUCTURE OF NUCLEIC ACIDS: A STRUCTURE FOR DEOXYRIBOSE NUCLEIC ACID*
CAMBRIDGE, UK

**ROSALIND FRANKLIN**
Franklin is one of the most controversial figures in science history. Born into a wealthy London family, she entered research soon after graduating, first in Cambridge and London during World War II, then in Paris and finally at King's. Her work on DNA was published in the same journal as the findings of Crick and Watson, but the two men received more acclaim than she and seldom highlighted her part in the discovery. In 1962, Crick and Watson received the Nobel Prize along with Maurice Wilkins (1916–2004), Franklin's boss at King's. Franklin had died from cancer by this time.

The chemical now known as DNA was discovered in 1869, and its chemical constituents of ribose sugar, phosphates and organic ringed molecules named nucleic acids were isolated in 1909. 'DNA' is short for deoxyribonucleic acid. Thereafter, DNA was shown to be the material involved in the inheritance of traits from one generation to the next. But no one knew how it did it. In order to find out, it was first necessary to determine how all its components fitted together. By the start of the 1950s, several research teams were working on this problem, taking two main approaches. At Cambridge, Francis Crick (1916–2004) and James Watson (1928–) tried to model the molecule on paper and with sticks and balls. At King's College, London, Rosalind Franklin (1920–1958) used X-ray crystallography to make images of the molecule to provide clues to its shape. Franklin interpreted one image, the famous Photo 51, as showing that DNA had a coiled or helical structure. This information was passed in secret to Crick and Watson, who used it to build an accurate model of the molecule in 1953. The ribose sugars chained together by the phosphates form structural backbones running along the sides of the molecule, while the nucleic acids pair up to form cross-connections. The inherited genetic code is set out in the order of these nucleic acids.

GENETICS **p.31** GENETIC MODIFICATION **p.38**

**Key publications by Francis Crick**

*Of Molecules and Men* 1966

*The Astonishing Hypothesis* 1994

**Key publication by James Watson**

*The Double Helix* 1968

James Watson (left) and Francis Crick (right) with their DNA double helix model (1953).

EVOLUTION BY NATURAL SELECTION **p.161** DNA PROFILING **p.205** CRISPR GENE EDITING TOOLS **p.206**

# Milgram Experiment

1961

**STANLEY MILGRAM:** *BEHAVIORAL STUDY OF OBEDIENCE* • NEW HAVEN, CONNECTICUT, USA

**Key publications by Stanley Milgram**

*The Experience of Living in Cities* 1970

*Television and Antisocial Behavior: Field Experiments* 1973

*Obedience to Authority* 1974

Carried out in 1961, this experiment conducted by Stanley Milgram (1933–1984) provides a fascinating insight into how people are obedient to authority. Volunteers were recruited from students at Yale University and they were paid in advance. One volunteer was tested at a time, only they were told that their role was as Teacher, and they followed instructions from the Experimenter, a scientist who was the authority figure in the room. The Teacher's job was to administer electric shocks to the Learner when he got simple tasks and questions wrong. The Experimenter explained that the shocks were painful but harmless. The Learner was unseen in another room, and the Teacher believed that the Learner was also a volunteer. However, the Learner was in fact one of the researchers, who gave out screams of pain when the Teacher administered the phoney shock. The Learner deliberately answered incorrectly from time to time, and the Experimenter asked the Teacher to increase the voltage in every instance. All the volunteers followed their instructions until the shock equipment was set at 300 volts. At this point, the Learner started giving cries of pain. About a third of volunteers stopped then, but the remaining two thirds carried on, even after the Learner had implored them to stop, grown weaker and then stopped responding as ever more powerful shocks were administered.

**STANLEY MILGRAM**

The son of Jewish immigrants, Milgram's early years in New York City were touched by news of family members falling victim to the Holocaust in Europe. He opted to study psychology so he could find out how people could behave in such inhuman ways. The obedience experiment that made his name was carried out just two years after finishing his PhD. Milgram also searched for a link between watching TV and bad behaviour, but found no such connection. He died of a heart attack at the age of 51.

NEUROSCIENCE AND PSYCHOLOGY p.34

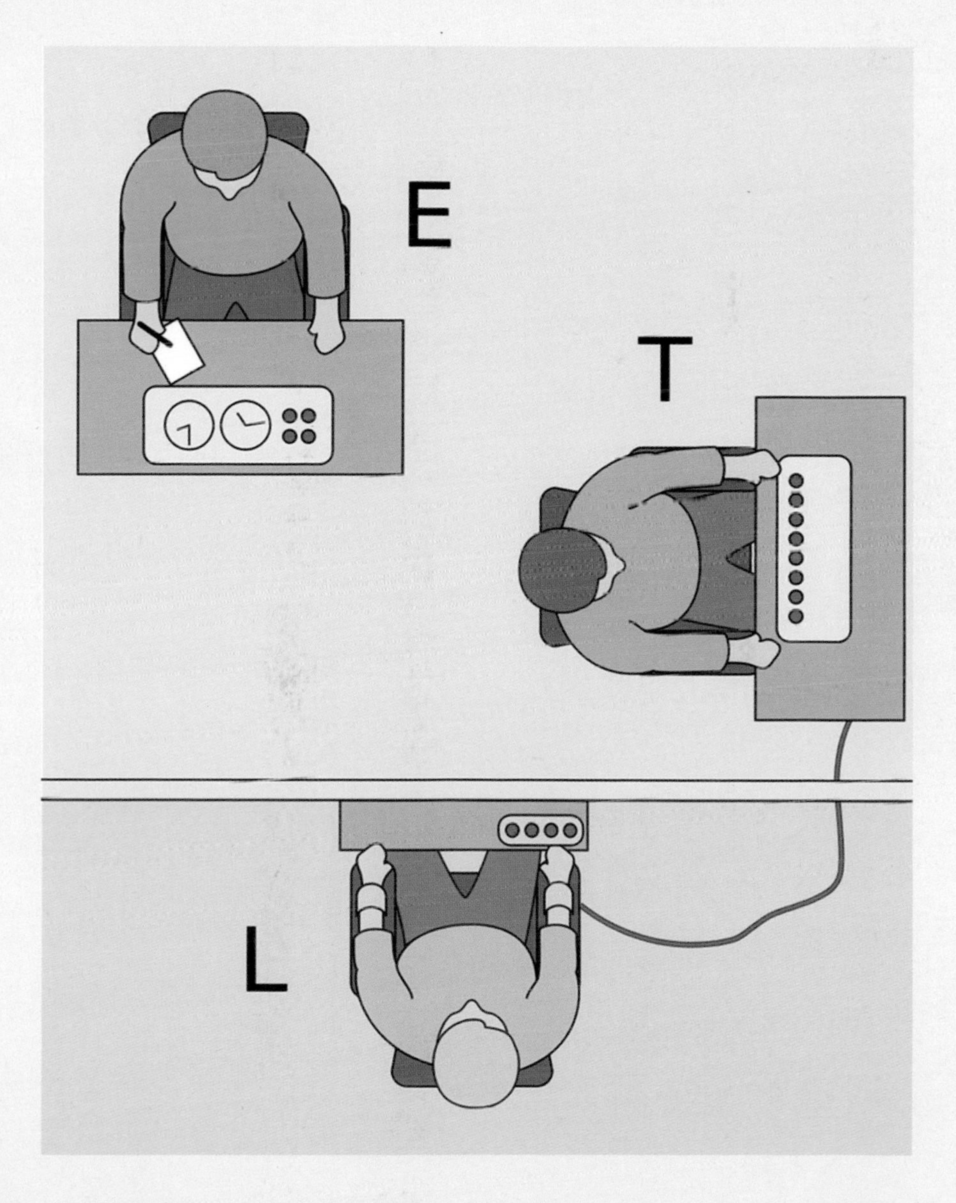


The experimenter (E) convinces the subject ('Teacher', T) to give what he believes are painful electric shocks to another subject, who is actually an actor ('Learner', L).

SCIENTIFIC PROCESS p.182 CLINICAL TRIALS p.208 MACHINE LEARNING p.213

# Cosmic Microwave Background

1965

**GEORGE SMOOT ET AL:** *WRINKLES IN TIME* • NEW YORK, USA

THE UNIVERSE IS MISSING **p.37**

**Key publication by George Smoot**
*Unity of Knowledge* 2002

**GEORGE F. SMOOT**
One of the lead researchers on COBE, George Smoot (1945–) won the 2006 Nobel Prize along with his collaborator John C. Mather (1946–). Smoot is reported to have given his prize money to charity. Three years later he appeared on *Are You Smarter than a 5th Grader?* on US TV and won $1 million. He continues to study the CMB along with dark energy and infrared astronomy.

For three decades there was no physical evidence for the Big Bang that was theorized to have started the Universe 13.8 billion years ago. However, in 1965 a faint glow of microwaves that filled every corner of the sky was discovered. This matched up with the Big Bang theory, which explained it was the first flash of light ever seen in the Universe, unleashed by the formation of the first atoms when the Universe was about 300,000 years old. The glow, which became known as the Cosmic Microwave Background (CMB), was discovered by accident while scientists were using a sensitive radio receiver to research satellite communication. The blast of light initially released has been steadily stretched by the expanding Universe into much lower-frequency microwaves, but it offers a snapshot of the fabric of the early Universe when matter and energy were much more densely packed. In 1989, the Cosmic Background Explorer (COBE) probe made a temperature map of the CMB. It is mostly very 'smooth' and consistently very cold, just 3.5 degrees Celsius above absolute zero. However, COBE found tiny inconsistencies in temperature, and subsequent probes have made even more detailed surveys. The warmer regions show where in space clusters of galaxies would form as the Universe expanded further, while the colder regions of the CMB are now great voids – enormous expanses of emptiness.

Before Smoot and Mather started work, the CMB was observed by chance in the 1960s by US astrophysicists Arno Penzias (left, 1933–) and Robert Woodrow Wilson (right, 1936–) through the horn radio antenna at the Bell Laboratories in Holmdel, New Jersey. In 1978 the two men were co-winners of the Nobel Prize in Physics.

BIG BANG **p.169** DARK MATTER **p.175** TELESCOPES **p.189**

# Exoplanets

1995

**NASA:** KEPLER MISSION

**Key publication**
Michael Perryman, *The Exoplanet Handbook* 2018

The idea that our Sun was the only star in the Universe with a solar system (a set of planets orbiting it) was very hard to believe. However, the process of detecting extrasolar planets (exoplanets) is not easy. Not only are they very small and faint when viewed from Earth, but also their light is completely lost in the glare of their star.

The first exoplanet orbiting a star was discovered in 1995 using a spectroscopic technique that detects changes in the colours of light from a star – changes that are indicative of the star being made to wobble slightly as its planets' gravity pulls on it. In 2009, NASA's Kepler observatory was launched into space to seek out exoplanets by a different method. When planets orbit in front of their star – a process known as 'transiting' – they block out some of the light, making the star dim. Away from the distortions caused by the atmosphere, Kepler's telescope stared at one patch of the heavens looking for stars that dimmed then brightened. Those that did might have Solar Systems, and spectroscopes back on Earth were trained on them to confirm the presence of exoplanets. The dimming effect and the amount of wobble were then used to calculate the size and orbits of the exoplanets. Were any like Earth? Kepler worked until 2018, scanning 530,506 stars and detecting 2,662 planets (with more possibles still to confirm). Several of them are small enough, and positioned in a habitable orbit, to be Earth-like and may harbour alien life.

**JOHANNES KEPLER**

**The exoplanet observatory was aptly named after this German astronomer. It was Kepler who formulated the laws of planetary motion (using data collected by other astronomers) to show that planets orbit their stars in ellipses, not in circles.**

ANCIENT ASTRONOMERS **p.12**

The Kepler spacecraft in a clean room at the Ball Aerospace & Technologies Corporation, Boulder, Colorado (2008).

ORIGIN OF THE SOLAR SYSTEM **p.179** TELESCOPES **p.189** PLANETARY ROVER **p.215**

# Discovery of Dark Energy

1998

**ADAM RIESS, BRIAN SCHMIDT AND SAUL PERLMUTTER:**
*OBSERVATIONAL EVIDENCE FROM SUPERNOVAE FOR AN ACCELERATING UNIVERSE AND A COSMOLOGICAL CONSTANT* • CERRO TOLOLO INTER-AMERICAN OBSERVATORY, CHILE; KECK OBSERVATORY, HAWAII, USA

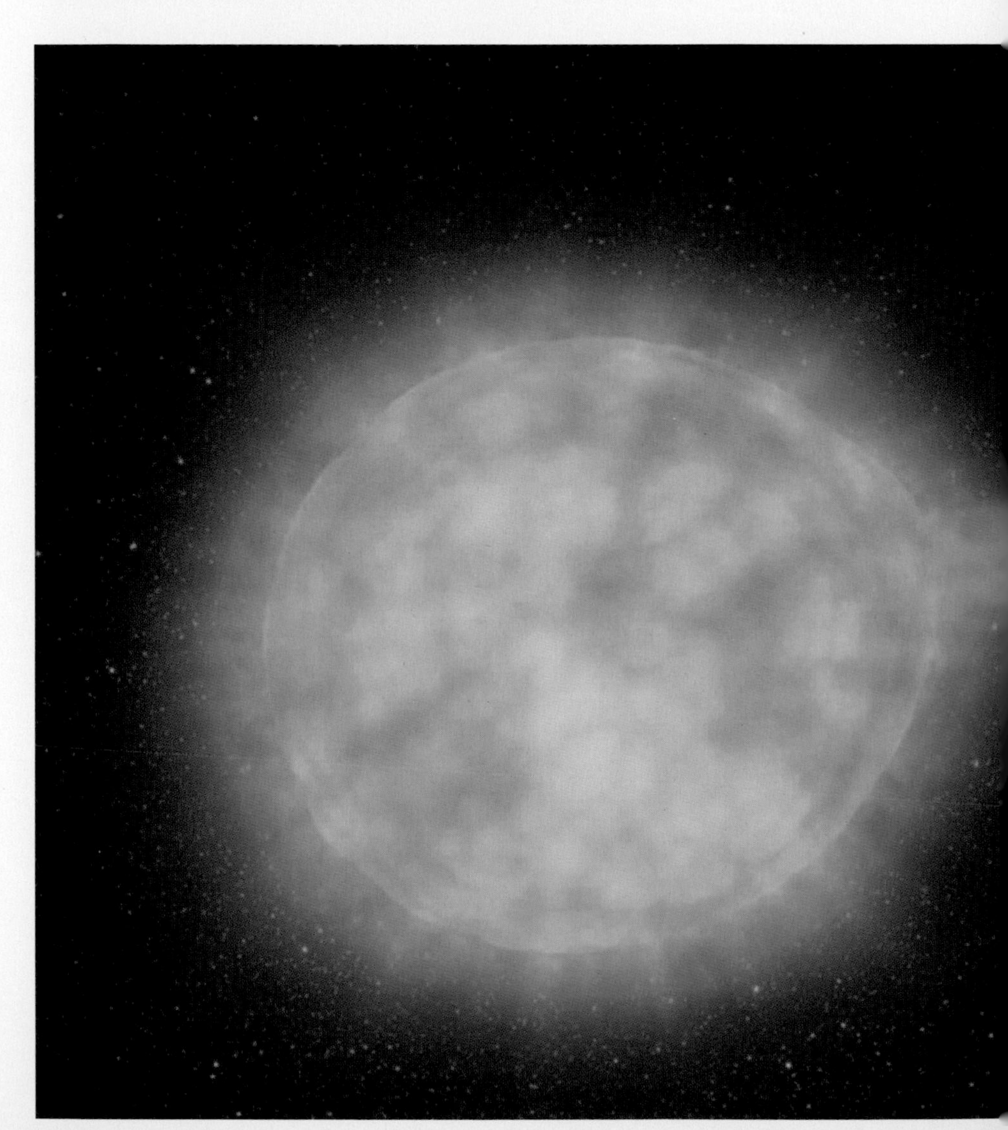

ANCIENT ASTRONOMERS **p.12** THE SIZE OF THE UNIVERSE **p.28** THE UNIVERSE IS MISSING **p.37**

**Key publication by Adam Riess**
*Physics for the Curious: Why Study Physics* 2016

**ADAM RIESS**

**Dark energy was discovered through two originally independent surveys, one run by Brian Schmidt (1967–), an Australian, along with Adam Riess (1969–) and another by Saul Perlmutter (1959–), who like Reiss is American. All three won the Nobel Prize in 2011. Riess had previously won multiple awards, including in 2008 a MacArthur 'Genius Grant' of $1 million.**

The discovery in 1929 that the Universe is expanding led in part to the formulation of the Big Bang theory, but it nevertheless left a big question unanswered: would the Universe continue to expand forever, or was the gravitational pull of all its stars applying the brakes, slowing the expansion and maybe putting it into reverse one day?

The answer depends somewhat on the mass of the Universe, including all the dark matter we cannot see. So in the 1990s surveys of the expansion began with reference to Type Ia supernovas – stars that explode when they reach exactly 1.44 times the size of our Sun. Knowing the size of these stars, astronomers can figure out how far away they are using their relative brightness – fainter ones being further away. The expansion of space also gives the stars' light a red shift due to the Doppler effect. It was predicted that comparing the expansion rates of older parts of the Universe with newer ones would show that the process was gradually slowing. In 1998, however, to general amazement, it was found that the opposite is happening: the expansion is speeding up. This is due to a mysterious anti-gravity effect known as dark energy, which is thought to be energy stored in a vacuum – the nothingness of space. And as the Universe expands, it makes more and more nothingness, which creates more expansion.

When a white dwarf reaches a mass equivalent to 1.44 times that of Earth's Sun it can no longer sustain its own weight, and so blows up, as shown here.

BIG BANG **p.169** DARK MATTER **p.175** COSMIC INFLATION **p.176**

# LIGO

2016

**KIP THORNE AND R.D. BLANDFORD:** *MODERN CLASSICAL PHYSICS*
PRINCETON, NEW JERSEY, USA

Einstein's 1915 general theory of relativity explained that objects warp, or curve, the space around them. One of the predictions from this theory was that objects ploughing through the Universe would leave a wake of ripples in space. It took 100 years to find them, but in 2016 the LIGO experiment located the ripples, or gravitational waves, created by the collision of two black holes in deep space. The waves are literally rarefactions and compressions in space itself, so measuring their effects is complicated by the fact that measuring devices are stretched and squeezed along with the space. LIGO stands for Laser Interferometer Gravitational-Wave Observatory, and as that name suggests it uses lasers to detect the waves. One laser is split into two perpendicular beams fired along 4-kilometre (2.5-mile) tunnels. At the far ends, mirrors reflect them back. One mirror is one half of a wavelength (a few billionths of a metre) further away than the other. So when the beams arrive back at the detector they are perfectly out of sync and cancel each other out. A passing gravitational wave stretches one of the tunnels and the beam of light inside, changing the latter's wavelength. As a result, the returning beams no longer cancel but combine into a tell-tale flickering laser signal of a gravitational wave. LIGO is now being enhanced to work as a telescope that can image the Universe by the gravity of its contents.

**KIP THORNE**

**A native of Utah, Kip Thorne (1940–) won the 2017 Nobel Prize for his work on LIGO, along with Rainer Weiss (1932–), who invented its laser interferometry system, and Barry Barish (1936–). Thorne founded the LIGO programme in 1984, and is also a world expert on black holes and wormholes. He was the scientific advisor for *Interstellar*, a 2014 sci-fi film about both these concepts.**

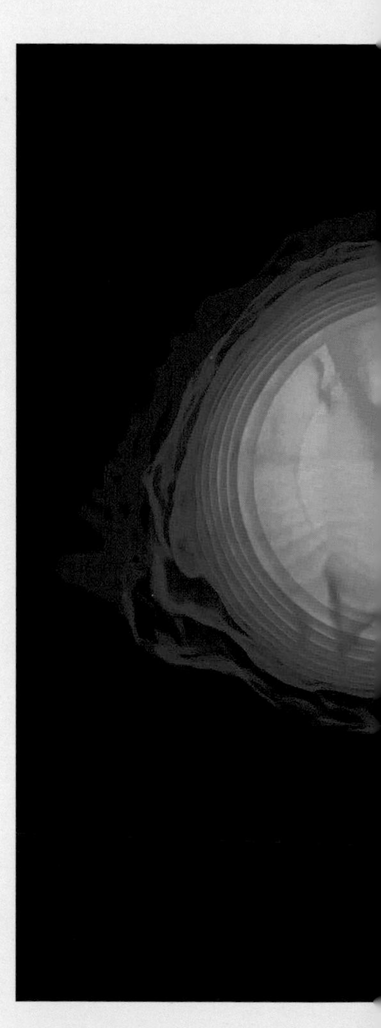

A pair of neutron stars colliding, merging and forming a black hole.

ANCIENT ASTRONOMERS **p.12** THE SIZE OF THE UNIVERSE **p.28**

**Key publications by Kip Thorne**

*Black Holes & Time Warps: Einstein's Outrageous Legacy* 1994

*The Science of Interstellar* 2014

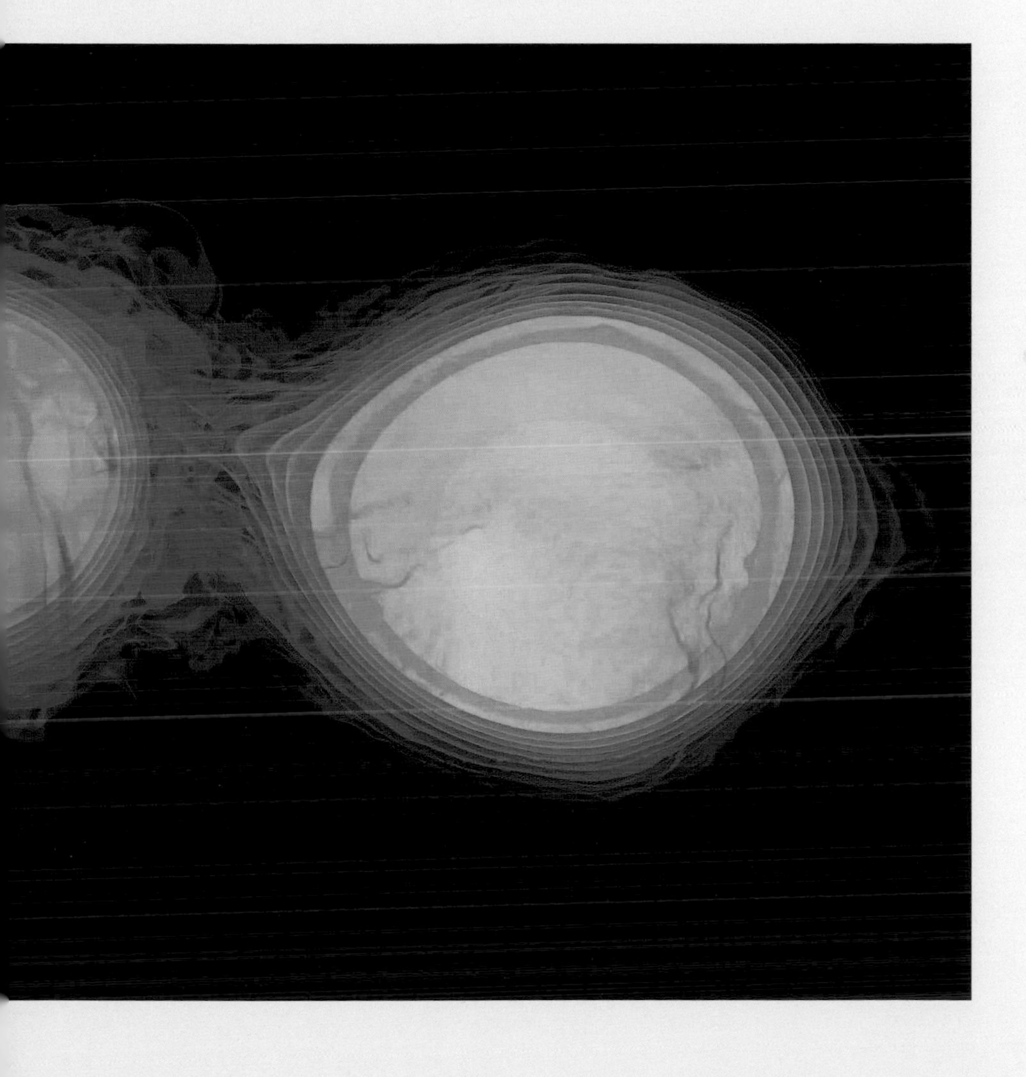

UNIVERSAL GRAVITATION **p.158** RELATIVITY **p.163** TELESCOPES **p.189**

THEORIES

# Panspermia

## KEY SCIENTISTS: ANAXAGORAS

What is the origin of life? The theory of panspermia proposes that the first life on Earth, or at least the complex chemical building blocks from which life arose, arrived from outer space. It is a surprisingly old idea, first proposed by Greek philosopher Anaxagoras (c.500–c.428 BCE). His general observation was that life was carried to Earth on flaming meteors that showered from the heavens. In essence, the theory has moved on little since then, but has been repeatedly fleshed out as our knowledge of biochemistry, genetics and their evolutions has developed. The foundation of the theory is that complex chemicals such as DNA could not form spontaneously, an idea strongly challenged by the Miller-Urey experiment (see page 140). However, the latest version of panspermia proposes that lithotrophs (rock-eating microbes) were delivered to Earth inside the frozen core of a meteorite, and survived the impact. This is not so far-fetched if one is happy to accept that cellular life simply emerged elsewhere, presumably in the Solar System, but how we do not know.

**KEY DEVELOPMENTS**

**While plants are autotrophs ('self-eaters' that create their own food supply) and animals are heterotrophs (eaters of other things), lithotrophs are relatively simple, single-celled microorganisms that use the iron, sulfur or nitrogen chemicals in rocks as a supply of energy. They are found at least 3 kilometres (2 miles) down in Earth's crust and may make up half the biomass of life on Earth.**

This 1783 mezzotint attempts to convey the spectacle of a meteor shower in the sky above Newark-on-Trent, England.

# Laws of Motion

**KEY SCIENTISTS:** ISAAC NEWTON • GALILEO GALILEI

In his 1687 masterwork *Philosophiæ Naturalis Principia Mathematica*, Isaac Newton presented a clockwork Universe in which all things, from light beams to planets, were orchestrated by three unbreakable Laws of Motion. The first law states that an object remains still or continues to move at a constant speed until a force acts on it to change that state of motion. The second uses the formula Force = Mass × Acceleration ($F=ma$) to relate how a force will accelerate an object of small mass more than it will accelerate a heavier object. (In other words, hitting something harder makes it move further.) The third and final law is the most quoted, stating something like, 'Every action has an equal and opposite reaction'. This phrase is widely and loosely interpreted today, but Newton meant that when a force pushes on an object, that object responds with an equal force pushing back in the other direction. That is why pushing on a heavy object may result in you going backwards, and the object staying still.

[ 1 ]

PHILOSOPHIÆ NATURALIS Principia MATHEMATICA.

Definitiones.

Def. I.

*Quantitas Materiæ eſt menſura ejuſdem orta ex illius Denſitate & Magnitudine conjunctim.*

AEr duplo denſior in duplo ſpatio quadruplus eſt. Idem intellige de Nive et Pulveribus per compreſſionem vel liquefactionem condenſatis. Et par eſt ratio corporum omnium, quæ per cauſas quaſcunq; diverſimode condenſantur. Medii interea, ſi quod fuerit, interſtitia partium libere pervadentis, hic nullam rationem habeo. Hanc autem quantitatem ſub nomine corporis vel Maſſæ in ſequentibus paſſim intelligo. Innoteſcit ea per corporis cujuſq; pondus. Nam ponderi proportionalem eſſe reperi per experimenta pendulorum accuratiſſime inſtituta, uti poſthac docebitur.

B Def.

The opening page of the first edition of Newton's most important work, known in English as *Mathematical Principles of Natural Philosophy*.

**KEY DEVELOPMENTS**

Motion is always relative to something else. On Earth that is generally the surface of the planet, which for our purposes is always regarded as still. Of course, relative to the Sun or the Galactic Centre, our planet is hurtling along at enormous speeds. It was Galileo who first expressed this notion years before Newton and later Einstein refined it, as it was found that the Newtonian Universe was at odds with other physical phenomena, not least light itself.

THE SCIENTIFIC REVOLUTION **p.18** PENDULUM LAW **p.52** ACCELERATION UNDER GRAVITY **p.55**
WEIGHT OF THE EARTH **p.74**

# Universal Gravitation

**KEY SCIENTISTS:** ISAAC NEWTON

As well as setting out the three Laws of Motion in his 1687 book, generally abbreviated to *Principia*, Isaac Newton also presented a complete theory of universal gravitation. This explained that the force that makes apples fall from trees is the same one that holds the Moon and planets in their orbits. Newton's innovation was to realize that the force of gravity acted between all objects, so the apple pulls on Earth and Earth pulls on the apple. Earth wins because its mass is so much greater, and the net effect is for the apple to move towards Earth. Additionally, the force is dependent on an inverse square relationship, which is to say that when the distance between two objects is doubled, the gravitational force acting between them is reduced by a factor of four. The influence of gravity fades fast: a ten-fold increase in distance reduces the force to just 1 per cent of the original. Newton used his Law of Universal Gravitation to explain the parabolic paths of projectiles, and showed how they could be fired into orbit or even escape Earth's gravity completely if moving fast enough.

**KEY DEVELOPMENTS**

Newton liked to keep the fruits of his work secret, often for many years, and this led to repeated disputes over priority, or who could lay claim to various discoveries. Famously Newton's feud with Gottfried Leibniz (1646–1716) over who invented calculus created a rift in the scientific establishment. With regard to gravity, it was Robert Hooke who dared to challenge Newton over the inverse square law. After Hooke died, Newton suppressed the work of his rival, who had been one of England's more prolific scientific researchers.

This portrait of Sir Isaac Newton, first formulator of the theory of gravitation, was painted between around 1715 and 1720.

THE RISE OF THE SCIENTIFIC INSTITUTION **p.19** THE SPACE RACE **p.32** ACCELERATION UNDER GRAVITY **p.55**
HOOKE'S LAW **p.62**

# Atomic Theory

**KEY SCIENTISTS:** DEMOCRITUS • JOHN DALTON

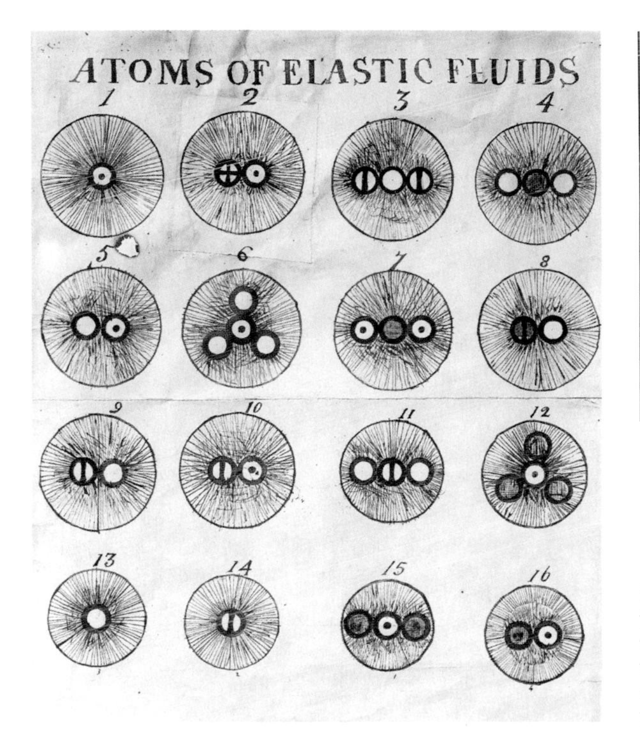


**KEY DEVELOPMENTS**

In 1808 John Dalton set out his atomic theory and built on it to explain the Law of Definite Proportions. Earlier chemists had noted that a particular weight of one element always combined with a particular weight of another to form a compound substance. Dalton proposed that this was due to the elements' atoms clustering together in orderly combinations, combinations that are today known as molecules.

One of a series of drawings of atomic formulae made in 1925 by John Dalton, whose theory was based on the concept that the atoms of different elements are distinguished by differences in their weights.

The word atom is derived from the ancient Greek for 'uncuttable', and the concept of a unit so small that it cannot be reduced comes from antiquity. It arose from a philosophical question about the nature of space and motion but was then used by Democritus (c.460–c.370 BCE) to explain the properties of substances. For example, some atoms seemed to be hooked so that they clumped together, while others were jagged and painful to the touch.

Atoms remained hypothetical entities until the early 1800s. Then John Dalton (1766–1844), whose principal interest lay in the link between air pressure and the weather, came across proof that gases behave as distinct substances even when mixed together, with each gas exerting a partial pressure that added up to a total. The explanation for this was that gases are made up of minute but uniquely distinct units, or perhaps particles or corpuscles – Dalton opted for the original term 'atom'.

ALCHEMY **p.15** THE BIRTH OF CHEMISTRY **p.20** GAS LAWS **p.60** BROWNIAN MOTION **p.90** DISCOVERY OF THE ELECTRON **p.114**

# Laws of Thermodynamics

**KEY SCIENTISTS:** JAMES PRESCOTT JOULE • LORD KELVIN • HERMANN VON HELMHOLTZ

Thermodynamics is the branch of physics that deals with the behaviour of energy. It is underwritten by three fundamental laws that were set out in the 1800s after many centuries of misdirection in which heat and light were regarded as material substances. Instead, energy is not a thing, more an ability to do work. To a physicist, work is applying a force over a given distance (and power is the rate at which that work is done). To do work, energy takes many forms, and this is the first law: energy cannot be created or destroyed; it can only be transformed into different forms, such as motion, heat, light or sound. The second law states that a closed, or discrete, system will increase in entropy over time. This complex statistical feature of energy transfer basically means that energy is always becoming more dispersed and diffuse. The third law states that the lowest possible temperature is absolute zero (0 Kelvin, or –273.15 degrees Celsius). It is impossible to reach this temperature because it would take an infinitely large refrigerator and an infinite amount of time.

**KEY DEVELOPMENTS**

**There is a zeroth law of thermodynamics, so numbered because it was well known before the others, but only added to the list after the first law was established. It states that two connected systems will enter thermal equilibrium, which more simply put means that a hot and a cold space will eventually both become the same temperature.**

William Thomson, 1st Baron Kelvin (1824–1907), whose temperature scale was adopted as an SI unit and now bears his name.

THE NEW PHYSICS **p.27** GAS LAWS **p.60** CARNOT CYCLE **p.85**

# Evolution by Natural Selection

**KEY SCIENTISTS:** CHARLES DARWIN • ALFRED RUSSEL WALLACE • JEAN-BAPTISTE LAMARCK

Theories of evolution are among the most significant contributions science makes to humans' search to understand themselves, and consequently among the most contentious. Evidence from rocks and fossils tells us that Earth is very old indeed, and the organisms that live here today are different from the ones in the past. Moreover, new forms of life developed, or evolved, from older ones. How is this possible? An answer to this question was offered by Charles Darwin in his 1859 book *On the Origin of Species*. Darwin saw that there is great variation among members of a species, with some being better suited to the challenges of survival than others. Nature selected them to live long and fertile lives, while those that did not fit their situation died without breeding. Natural selection ensures that successful features are passed down the generations, and that anything unable to compete is edited out. Variation is not stagnant, and new features are always appearing, so natural selection is always at work, gradually altering all life on Earth.

**KEY DEVELOPMENTS**

Darwin's theories did not arise out of nowhere. His own grandfather, Erasmus Darwin (1731–1802), had noticed how the drive for survival favoured the fittest. Erasmus's contemporary, Jean-Baptiste Lamarck (1744–1829), proposed that evolution was driven by the acquisition of characteristics in life, such as the calloused hands and big muscles of blacksmiths, which were then passed on to their children. Charles Darwin's theory was near duplicated by Alfred Russel Wallace (1823–1913), and the pair jointly published some of their ideas.

The idea that humans were related to apes was not universally welcomed at first, as this 1874 cartoon attests.

EXISTENCE OF GENES **p.106** THE FUNCTION OF CHROMOSOMES **p.109** ORIGIN OF LIFE **p.140** THE DOUBLE HELIX **p.142**

# The Periodic Table

**KEY SCIENTISTS:** DMITRI MENDELEEV

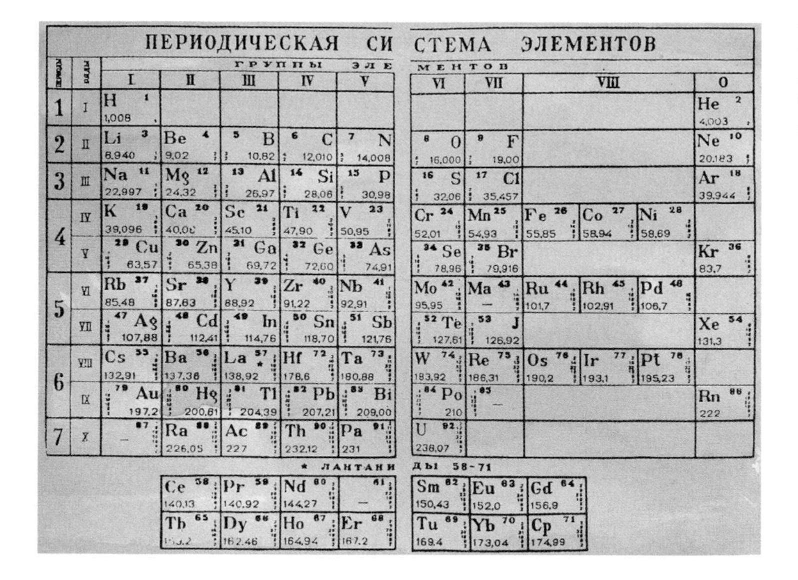

ПЕРИОДИЧЕСКАЯ СИСТЕМА ЭЛЕМЕНТОВ

ГРУППЫ ЭЛЕМЕНТОВ

| | | I | II | III | IV | V | VI | VII | VIII | | | 0 |
|---|---|---|---|---|---|---|---|---|---|---|---|---|
| 1 | I | H 1 1,008 | | | | | | | | | | He 2 4,003 |
| 2 | II | Li 3 6,940 | Be 4 9,02 | 5 B 10,82 | 6 C 12,010 | 7 N 14,008 | 8 O 16,000 | 9 F 19,00 | | | | Ne 10 20,183 |
| 3 | III | Na 11 22,997 | Mg 12 24,32 | 13 Al 26,97 | 14 Si 28,06 | 15 P 30,98 | 16 S 32,06 | 17 Cl 35,457 | | | | Ar 18 39,944 |
| 4 | IV | K 19 39,096 | Ca 20 40,08 | Sc 21 45,10 | Ti 22 47,90 | V 23 50,95 | Cr 24 52,01 | Mn 25 54,93 | Fe 26 55,85 | Co 27 58,94 | Ni 28 58,69 | |
| | V | 29 Cu 63,57 | 30 Zn 65,38 | 31 Ga 69,72 | 32 Ge 72,60 | 33 As 74,91 | 34 Se 78,96 | 35 Br 79,916 | | | | Kr 36 83,7 |
| 5 | VI | Rb 37 85,48 | Sr 38 87,63 | Y 39 88,92 | Zr 40 91,22 | Nb 41 92,91 | Mo 42 95,95 | Ma 43 — | Ru 44 101,7 | Rh 45 102,91 | Pd 46 106,7 | |
| | VII | 47 Ag 107,88 | 48 Cd 112,41 | 49 In 114,76 | 50 Sn 118,70 | 51 Sb 121,76 | 52 Te 127,61 | 53 J 126,92 | | | | Xe 54 131,3 |
| 6 | VIII | Cs 55 132,91 | Ba 56 137,36 | La 57 * 138,92 | Hf 72 178,6 | Ta 73 180,88 | W 74 183,92 | Re 75 186,31 | Os 76 190,2 | Ir 77 193,1 | Pt 78 195,23 | |
| | IX | 79 Au 197,2 | 80 Hg 200,61 | 81 Tl 204,39 | 82 Pb 207,21 | 83 Bi 209,00 | 84 Po 210 | 85 — | | | | Rn 86 222 |
| 7 | X | 87 — | Ra 88 226,05 | Ac 89 227 | Th 90 232,12 | Pa 91 231 | U 92 238,07 | | | | | |

* ЛАНТАНИДЫ 58-71

| | | | | | | |
|---|---|---|---|---|---|---|
| Ce 58 140,13 | Pr 59 140,92 | Nd 60 144,27 | 61 — | Sm 62 150,43 | Eu 63 152,0 | Gd 64 156,9 |
| Tb 65 [illegible] | Dy 66 162,46 | Ho 67 164,94 | Er 68 167,2 | Tu 69 169,4 | Yb 70 173,04 | Cp 71 174,99 |

The Periodic Table of the Elements in Russian Cyrillic script (1925).

The Periodic Table is a graphical representation of all the elements known to science. It offers information at a glance by grouping elements with similar chemical properties together. The table's format mirrors the differences in each element's unique subatomic structure, yet it was established by Dmitri Mendeleev (1834–1907) in 1869 before he (or anyone else) knew about how the atom was constructed from electrons, protons and neutrons. Beginning with the lightest, hydrogen, Mendeleev set out the elements in order of increasing atomic mass (the weight of each atom). He also used a repeating, or periodic, pattern in the elements' combining power. As usually laid out, the Periodic Table has 7 rows (periods) and 18 columns (groups). Elements in the same period have similar-sized atoms, energy and electronic properties. Elements in the same group generally have the same electron configurations.

**KEY DEVELOPMENTS**

**Mendeleev's table proved its worth when the Russian chemist used it to predict the properties of as yet undiscovered elements – and got them almost exactly right. Nearly 50 years later it was revealed why when the link to subatomic structure was revealed. The table's periods equate to electron shells around the atom. When one is full of electrons, a new shell and a new period are started.**

THE BIRTH OF CHEMISTRY **p.20** DISCOVERY OF RADIOACTIVITY **p.112** DISCOVERY OF THE ELECTRON **p.114** ATOMIC NUCLEUS **p.126** ATOMIC THEORY **p.159**

# Relativity

**KEY SCIENTISTS:** ALBERT EINSTEIN

**KEY DEVELOPMENTS**
Einstein figured out special relativity in 1905. By 1916 he had generalized it, creating the first of the two theories that still underpin physics (the other is quantum mechanics). General relativity includes a theory of gravity: all energy, of which mass is a very condensed form, warps space. As a second mass enters this warped space, its straight path is bent towards the first mass and so it appears to fall towards it.

Albert Einstein in 1921, the year he was awarded the Nobel Prize in Physics.

At the start of the twentieth century, some senior physicists believed that they had more or less figured out how the Universe worked. Any disparities between theory and observation were due to human errors and inaccurate devices. However, Albert Einstein, an amateur physicist, was not convinced. He wanted to know how it was possible that the speed of light was always constant when measured, when speed was relative in a Universe where everything was meant to move according to Newton's laws. Why then did the beam of light on the front of an approaching train arrive at an observer at the same time as the beam from a stationary light beside the track? In Newton's Universe, the train light should be moving faster. Einstein's answer was the theory of special relativity, according to which a mass (the train, for example) moving faster through space will move slower through time. The train's passage through time has slowed just enough for its light not to break the Universe's speed limit – the speed of light.

THE NEW PHYSICS **p.27** THE UNIVERSE IS MISSING **p.37** BROWNIAN MOTION **p.90** THE NONEXISTENCE OF ETHER **p.108** DISCOVERY OF ELECTROMAGNETIC WAVES **p.110** LIGO **p.152** LAWS OF MOTION **p.157**

# Plate Tectonics

**KEY SCIENTISTS:** ALFRED WEGENER • MARIE THARP

Marie Tharp at her draughting table in Lamont Hall, Lamont-Doherty Earth Observatory, Columbia University, USA, around 1961.

When the first accurate world maps were produced in the late sixteenth century, geographers began to comment on the way the continents appeared to be pieces of a big jigsaw, which would fit together if not separated by the oceans. Perhaps they had once been connected before being split apart by the forces of Earth? This hunch was given credibility by the work of Alfred Wegener (1880–1930) in the early 1900s. He showed that rock formations on opposite sides of the Atlantic were identical in age and make-up. They formed at the same time, in the same place. It took another 40 years to flesh out this idea of 'continental drift' into a theory of plate tectonics: Earth's crust is cracked into dozens of sections or plates. The plates float on the churning molten interior of the planet's mantle. At some of the boundaries between the plates, molten rock (magma) thrusts upwards, forcing the plates apart. The magma fills the gap and forms new crust. At other 'destructive' boundaries, one plate sinks beneath another, melting away into the mantle.

**KEY DEVELOPMENTS**

**A big step forward in the formulation of the field of plate tectonics – 'tectonics' means 'concerning building' – was the discovery of the Mid-Atlantic Ridge in 1953. Geologist Marie Tharp (1920–2006) used echo-soundings of the Atlantic to map out the sea-floor and thus reveal a vast mountain range running north to south down the middle. This ridge was later identified as a significant constructive boundary, where plates are being pushed apart.**

GEOLOGY AND THE EARTH SCIENCES **p.23** THE SIZE OF EARTH **p.44**

# Four Fundamental Forces

**KEY SCIENTISTS:** ISAAC NEWTON • ALBERT EINSTEIN • MARIE CURIE • MURRAY GELL-MANN

Physics reduces all activity in the Universe to the effects of four fundamental interactions, or forces. The most familiar is gravity. This is the weakest force, but it acts over the greatest distances – essentially the unlimited span of the Universe. Gravity is a force of attraction applied by all masses, and its strength is proportional to mass. The gravity of massive objects like black holes adds up to a significant pull. Nevertheless, the next force, electro-magnetism, is 10 billion trillion times stronger than gravity, although seldom applied in such magnitude. It is characterized by the phrase 'opposites attract and likes repel'. It is behind magnetic forces and electrical current, but at its root electromagnetism is the force that holds negatively charged electrons and positively charged protons together in an atom and maintains chemical bonds. The remaining two forces, the strong and weak interactions, only act within the minute confines of atomic nuclei. The strong interaction (the strongest force of all) holds it all together, locking in the protons and neutrons (and the quark particles within), while the weak interaction pushes out particles as radioactivity when the nucleus is inherently unstable. In 2021 physicists working near Chicago found that muons (subatomic particles like heavy electrons) were behaving in a way that could not be explained by the four fundamental forces. This result suggested that there is a fifth force of nature but more work is needed to say whether this is actually true.

**KEY DEVELOPMENTS**

**Electromagnetism and the strong and weak interactions are explained by the Grand Unified Theory (GUT), which sets out how all three forces were as one during the very early stages of the Universe. The temperature was so high that the three acted as the electronuclear force for the first $10^{-36}$ (0.000000000000000000000000000000000001) seconds. Then the strong force separated, but the GUT explains that electromagnetism and the weak interactions are still unified despite acting separately at low energies. The search is still on to find how to include gravity in the GUT.**

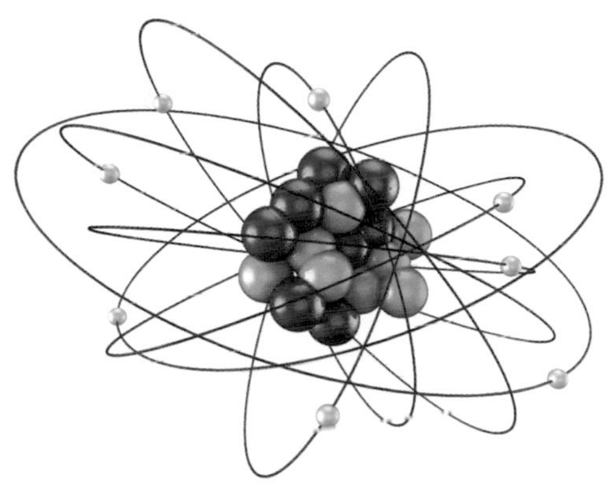

Diagram of one atom of oxygen with eight electrons orbiting eight protons and eight neutrons.

THE STANDARD MODEL **p.174** GEIGER-MÜLLER TUBE **p.191** PARTICLE ACCELERATORS **p.199**

# Uncertainty Principle

**KEY SCIENTISTS:** WERNER HEISENBERG • LOUIS DE BROGLIE • MAX BORN

The Uncertainty Principle is most associated with Werner Heisenberg (photographed here in the late 1960s), and is often referred to as Heisenberg's Uncertainty Principle. Heisenberg was a leading figure in developing quantum mechanics, which uses mathematical tools to explain and understand the subatomic world.

As if the idea that light could work like a wave and like a particle was not confusing enough, in 1924 French theoretical physicist Louis de Broglie (1892–1987) suggested that it was not just photons, the particles of light, that behaved like this, but all subatomic particles. It was already understood that the energy carried by a light wave was encoded into its frequency and wavelength, but de Broglie's proposal was to treat all particles as 'waveforms' that represented energy, position and other physical properties in the same way.

In 1927, George Thomson (son of J.J. Thomson) proved that electrons did indeed also behave as waves. At the same time, two German physicists, Werner Heisenberg (1901–1976) and Max Born (1882–1970), found that waveforms only worked when probability (the mathematics of chance) was included. This meant that if a particle's position was known, then its motion could only be estimated (and vice versa). The more exact the position, the more uncertain the estimate of motion. This idea, which relates to other quantum properties, became the Uncertainty Principle.

**KEY DEVELOPMENTS**

**The Uncertainty Principle is what gives quantum particles the spooky ability to be more than one thing at once, a feature known as superposition. A particle can literally be in more than one place at a time, and only once measured does the waveform 'collapse' and settle into a definite location. The uncertainty of quantum mechanics disrupts the link between cause and effect, because one cause can have multiple effects, and only chance will dictate which one occurs.**

THE NEW PHYSICS **p.27** DOUBLE SLIT EXPERIMENT **p.86** SPECTROSCOPY **p.102** DISCOVERY OF ELECTROMAGNETIC WAVES **p.110** DISCOVERY OF THE ELECTRON **p.114** WAVE-PARTICLE DUALITY **p.128**

# Quantum Physics

**KEY SCIENTISTS:** MAX PLANCK • ALBERT EINSTEIN • NIELS BOHR

Quantum physics arose from investigation of electrons and protons. In the early 1900s, Max Planck (1858–1947) found that the only way he could explain how objects radiated out their energy (as light, heat and other radiation types) was if they released it in tiny packets, or quanta. Quanta are not uniform and come in various magnitudes, but it is impossible to change the size of a quantum by removing or adding energy. In 1905, Einstein postulated that quanta were carried by particles named photons, and that light behaved both as a wave and as a stream of particles. (The energy in each photon dictated the wavelength, or colour, of the light.) This idea was built on by Niels Bohr (1885–1962), who found that atoms could only absorb and release packets of energy of a certain size. This explains why atoms of a particular element always release a unique set, or spectrum, of coloured lights (plus invisible radiation, too).

**KEY DEVELOPMENTS**

Bohr's model of how the atom works introduced the idea of electron orbitals. These are energy levels surrounding an atom's nucleus within which a certain number of electrons with a certain energy can sit. The precise energies are specific to each kind of atom. When a photon carrying the right amount of energy hits the atom, it is absorbed and makes an electron perform a 'quantum leap' to a more energetic orbital. As it drops back to its original position, the electron releases the energy again in the form of another photon.

Niels Bohr, shown here in 1923, was the proponent of the Copenhagen interpretation of quantum physics, according to which every specific physical property appears by chance from a set of alternatives.

THE SIZE OF THE UNIVERSE **p.28** DOUBLE SLIT EXPERIMENT **p.86** SPECTROSCOPY **p.102**
DISCOVERY OF THE ELECTRON **p.114** MEASURING CHARGE **p.120** WAVE-PARTICLE DUALITY **p.128**

# Valence Bond Theory

**KEY SCIENTISTS:** LINUS PAULING

In their natural state, nearly all atoms are connected, or bonded, to one or more other atoms to form molecules. Elements are sometimes pure, but more often they combine with one or more other elements to create compounds, such as water (hydrogen and oxygen). Most elements have atoms where the outermost orbital is incomplete – there is space for more electrons. Atoms form bonds in order to complete this outer orbital and achieve a more stable state than when unbonded. (Only the noble gases, such as helium, have full outer orbitals, and so they are almost entirely chemically inert and do not form bonds.) Of the 10 million compounds currently known, more than 90 per cent use covalent bonding, in which neighbouring atoms share electrons and thus fill up their outer orbitals. This concept evolved in the early twentieth century, mainly but not exclusively through 'On the Nature of the Chemical Bond', the landmark 1931 paper by US scientist Linus Pauling (1901–1994).

**KEY DEVELOPMENTS**

**Atoms are also linked by metallic and ionic bonds. Metallic elements always have just a few outer electrons, and so will lose them easily. In an ionic bond, a metal has lost its outer electrons and transforms into a positive ion. The electrons are donated to a non-metal, which becomes a negative ion, and these oppositely charged ions are attracted together. In pure and alloyed (mixed) metals, metallic bonds form as the outer electrons break free from individual atoms to form a sea of shared electrons which hold the atoms together.**

Linus Pauling with a molecular model, his formulation of which won him the first of his two Nobel Prizes in Chemistry (1954 and 1962).

THE BIRTH OF CHEMISTRY **p.20** ELECTRICITY **p.24** OXYGEN **p.72** ATOMIC THEORY **p.159**

# Big Bang

**KEY SCIENTISTS:** GEORGES LEMAÎTRE

**KEY DEVELOPMENTS**

The Big Bang is not a static single theory but an umbrella term for a collection of ideas that explain how the Universe came to be the way it is today. While the theory of Big Bang nucleosynthesis – which sets out the events of the early Universe that led to atoms, stars and galaxies – has been a success, the earliest moments of the Universe remain one of several unsolved mysteries, and the theory offers no insight into how time, space and energy formed.

Monseigneur Georges Lemaître was the first scientist to theorize the notion of the expanding Universe.

Analysis of Einstein's theory of relativity revealed that it was impossible for the Universe to be static. The perfect clockwork Universe as imagined by Newton and many others before him, in which nothing ever changed, was a myth. Instead, the Universe had to be either getting bigger or getting smaller. In 1927, Georges Lemaître (1894–1966), a Belgian professor of physics and a Catholic priest, opted for the former (and its expansion was confirmed in 1929). If the Universe was getting bigger, then it must have been smaller in the past and at some point in the distant past occupied a single dimensionless point in space.

Lemaître called it the Cosmic Egg, but work in the late 1940s explained how a super-hot Universe about the size of a grapefruit expanded and cooled and in so doing formed, step by step, the Universe's atomic matter. One of the theory's detractors attempted to belittle it as 'some kind of Big Bang', but that name stuck, as has the theory.

THE SIZE OF THE UNIVERSE **p.28**

# Stellar Nucleosynthesis

**KEY SCIENTISTS:** RALPH ALPHER • GEORGE GAMOW • ARTHUR EDDINGTON • FRED HOYLE

Composite false-colour photograph taken by three NASA observatories of Cassiopeia A, a supernova 10,000 light years from Earth.

The adage 'We are stardust' may be rooted in the alternative scene of the 1960s, but it also represents a general truth about the formation of the elements which was entering public understanding around this time, as set out in stellar nucleosynthesis, a theory of how the many elements we see in the Universe were formed.

There are about 90 elements occurring naturally on Earth, plus a few more in high-energy events in deep space. Hydrogen atoms make up 75 per cent of the matter in the Universe. They were formed soon after the Big Bang. Helium atoms make up another 23 per cent. The common elements – the likes of carbon, oxygen and iron – are formed by the fusion of smaller atoms inside stars. Stars are vast balls of hot hydrogen plasma, and in their cores the pressure is so immense that the hydrogen fuses into helium. That fusion releases more energy, which eventually shines out on the surface as heat and light. When the hydrogen runs out, older stars fuse helium into heavier elements, which eventually form clouds of gas and dust from which new stars and their planets, moons and perhaps life will one day form.

**KEY DEVELOPMENTS**

**Most stars, including our Sun, are dwarfs – neither big nor powerful enough to make elements heavier than iron. After burning helium as a red giant, a star simply diffuses into a nebulous cloud (leaving behind a hot core, known as a white dwarf). However, supergiant stars go out with a bang – a supernova – which is so violent it can synthesize much heavier and rarer elements, such as gold, uranium and xenon.**

SPECTROSCOPY **p.102** COSMIC RAYS **p.124** BIG BANG **p.169**

# Lock-and-Key Theory

**KEY SCIENTISTS:** EMIL FISCHER

The lock-and-key theory proposes a mechanism for the action of enzymes – biological catalysts that facilitate reactions which would otherwise not occur or do so only very slowly. All life uses enzymes. They are at work in all cells and in secretions such as saliva and stomach juices.

The theory was set out in 1894 by German chemist Emil Fischer (1852–1919), who imagined how the enzyme and its target chemical click together to make a temporary structure. All enzymes are made from proteins, which are polymers folded into a unique shape. Part of that shape is the active site (the lock). The materials processed by the enzyme, known as the substrate, fit precisely into the lock, forming the key. Once held by the active site, the substrate can react, causing one molecule to break apart or two to join together.

The first known enzyme was diastase, which was isolated in 1833 as the active chemical in the digestion of starch into simpler sugars. More than 5,000 enzymes have now been identified.

**KEY DEVELOPMENTS**

**The shape of a protein is crucial to its function. Proteins are polymers – chemicals made from chains of smaller molecules named monomers. Among the monomers are amino acids, of which there are about 20 used by life. The precise order of amino acids – the so-called primary structure of the protein – determines how the molecule will fold and refold on itself to create the perfect shape for its enzymatic role.**

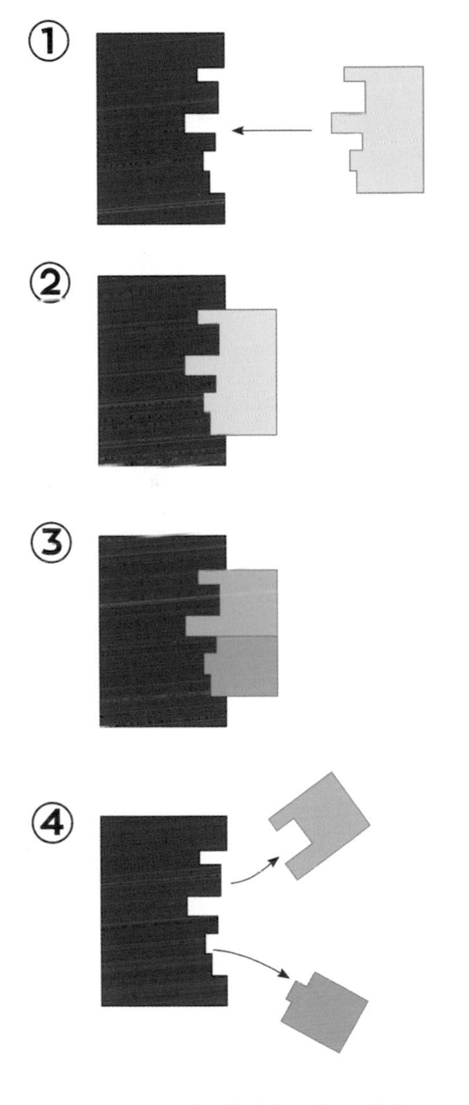


Diagrammatic representation of the lock-and-key mechanism of enzymes: (1) Substrate moves towards active site of enzyme, (2) Substrate starts reacting; (3) Reaction complete, two products formed; (4) Products leave active site.

CELL THEORY **p.25** CITRIC ACID CYCLE **p.139** ORIGIN OF LIFE **p.140**

# Central Dogma of Biology

**KEY SCIENTISTS:** FRANCIS CRICK • JAMES WATSON

This grandly named concept attempts to explain the mechanisms by which genetic sequence information is transmitted in biological entities. First outlined in the late 1950s by Francis Crick, the basic idea is that, while such data can be transferred from one nucleic acid to another and then converted into protein, once that conversion is completed the information can no longer change its form or its position.

The Central Dogma grew from the discovery of the double helix structure of deoxyribonucleic acid (DNA). It divides the possible modes of information transmission into three groups, the most important of which comprises the replication of DNA; this is the usual form of biological information transfer. DNA can also be copied into messenger ribonucleic acid (mRNA) and subsequently translated back into DNA. But a protein's shape can never be transmitted as RNA back to the nucleus.

**KEY DEVELOPMENTS**

**Ribonucleic acid (RNA) plays a significant role in the translation of genes into proteins. The genetic code travels from the chromosomes as mRNA to the ribosomes, which are themselves made of tangled units of ribosomal RNA (rRNA). The mRNA slides through the ribosome, and each three-letter unit, known as a codon, is matched up with a corresponding piece of transfer RNA (tRNA). It is tRNA's job to haul the right amino acid into line, before the mRNA moves forward, and the next codon signals which acid is needed next.**

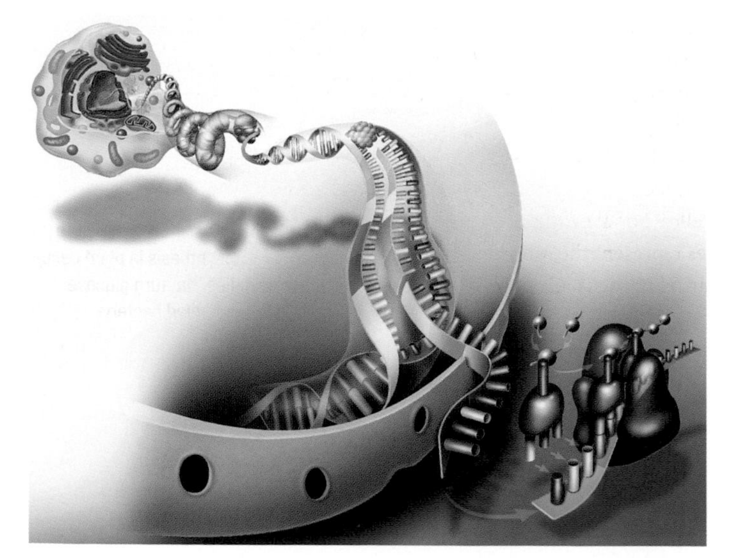

This diagram illustrates how the DNA double helix can be split by RNA. The code held by DNA carries information on how to build chemicals with a specific structure that are needed for life.

CELL THEORY **p.25** GENETICS **p.31** THE DOUBLE HELIX **p.142** LOCK-AND-KEY THEORY **p.171**

# Endosymbiosis

**KEY SCIENTIST:** LYNN MARGULIS

Through her work as a researcher and educator, American biologist Lynn Margulis raised public awareness of this field so much that she has been described as having done for symbiosis what Charles Darwin did for the theory of evolution.

Life on Earth may be divided into two groups, prokaryotes and eukaryotes. The former comprise bacteria and a similarly microscopic domain of life known as the archaea. These organisms have small cells without obvious internal structures. The eukaryotes include everything else, from algae and amoebas to oak trees and humans. Eukaryotic cells are ten times larger than prokaryote cells and contain a nucleus plus a set of complex internal structures known as organelles. Fossil evidence shows that prokaryotes represent the more primitive form, appearing in rocks 3.5 billion years ago, a full 2 billion years before the eukaryotes. The theory of endosymbiosis is the brainchild of American biologist Lynn Margulis (1938–2011). In 1967 she proposed that eukaryotes evolved from groups of unrelated prokaryotes living together in symbiosis and that an archaeon evolved a convoluted cell membrane after benefiting from an increased surface area. Then bacteria entered this host cell by chance and formed the first eukaryote. Whether this event occurred more than once is not known; perhaps not, because all today's eukaryotic organisms are descended from a single cell.

**KEY DEVELOPMENTS**

**The ideas behind the theory of endosymbiosis began to emerge as microscopy developed the capacity to show organelles in greater detail. As early as the 1910s, people remarked how the chloroplasts (sites of photosynthesis in plant cells) and mitochondria (organelles that turn glucose into energy for the cell) resembled bacteria. Mitochondria even have their own genetic code that is separate from the chromosomes, and analysis of that DNA suggests these organelles were originally free-living purple bacteria, still a common type, while chloroplasts evolved from cyanobacteria (also known as blue-green algae).**

CELL THEORY **p.25** EVOLUTION BY NATURAL SELECTION **p.161**

# The Standard Model

**KEY SCIENTISTS:** J.J. THOMSON • MURRAY GELL-MANN • PETER HIGGS

The Standard Model is a name for the set of subatomic particles that form the Universe and drive all its processes. The current model includes 18 particles, which are divided up in different ways. The principal division is that between fermions and bosons. Fermions are the particles involved in making ordinary matter, such as atoms. They are themselves split into quarks and leptons. Quarks occur in combination with other quarks or with antiquarks to form hadrons. Leptons are negatively charged and include electrons. The particles that transfer energy between fermions are known as bosons.

The Standard Model accurately predicts the interactions of quarks and leptons, but cannot foretell the masses of these particles or the strengths of their interactions. Yet scientists persist with it, in spite of its limitations, in the hope that it may gradually be developed into a complete unified theory that encompasses all subatomic particles: the strong, the weak and the electromagnetic.

**KEY DEVELOPMENTS**

**For many years, all known bosons were force carriers. But then in 2012 a new boson was discovered – the Higgs boson, which is not a force carrier, but instead gives fermions their mass. Additionally, every particle in the Standard Model has an antiparticle which is identical in size but opposite in charge. When a particle and an antiparticle meet, they annihilate each other.**

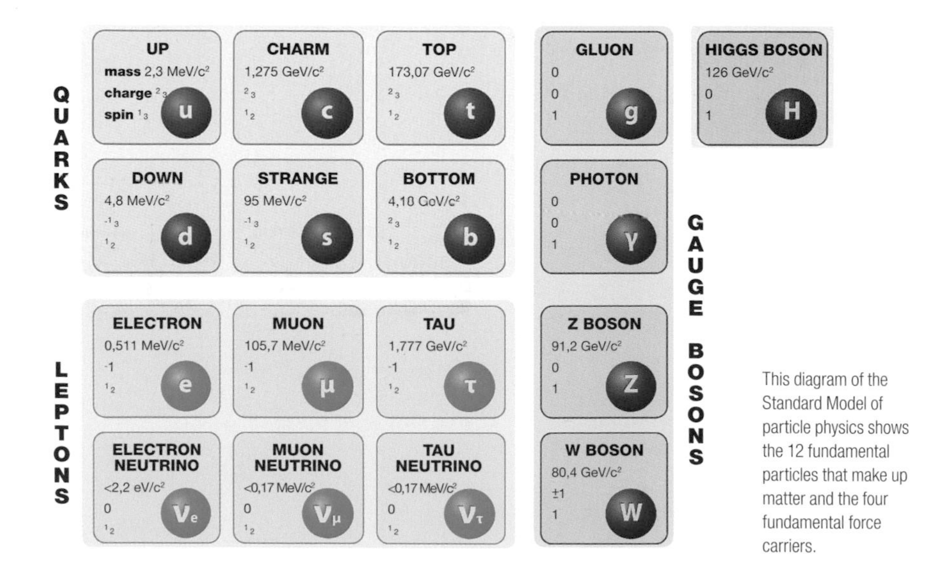


This diagram of the Standard Model of particle physics shows the 12 fundamental particles that make up matter and the four fundamental force carriers.

# Dark Matter

**KEY SCIENTISTS:** FRITZ ZWICKY • VERA RUBIN

By the 1930s it was clear that the Solar System was part of a galaxy named the Milky Way, which itself was separated from other galaxies by nothing but emptiness. As astronomers began to investigate the size and shape of the galaxy, they found that it seemed to be rotating faster than its apparent weight allowed. If the weight of the galaxy was based solely on the number of its stars, then the stars should be flung out in all directions as the galaxy turned at speed. Most astronomers assumed the observational data were wrong. Nevertheless, Swiss American Fritz Zwicky (1898–1974) suggested there might be some invisible material in the galaxy – '*dunkle Materie*' he called it, which is translated as dark matter.

The problem was then ignored until the 1970s, when American astronomer Vera Rubin (1928–2016) announced that her painstakingly precise measurements of the rotation of Andromeda, a big galactic neighbour, revealed that galaxies did indeed contain dark matter, about six times as much the visible matter. This ratio holds today; most of the Universe's matter is missing.

**KEY DEVELOPMENTS**
Dark matter can only be detected by its gravitational effects, and since it does not interact with light or any other radiation, it is invisible. There are two proposed sources: massive compact halo objects (MACHOs) or dense and dim material in the periphery of galaxies easily missed in observations. Yet neither is likely to account for much of the missing matter. Perhaps more persuasive is the claim of weakly interacting massive particles (WIMPs), which are proposed to be everywhere but remain undetected.

The Andromeda Galaxy photographed from the Galaxy Evolution Explorer in 2003 (the spacecraft was finally decommissioned in 2013).

 THE UNIVERSE IS MISSING **p.37** DISCOVERY OF DARK ENERGY **p.150**

# Cosmic Inflation

**KEY SCIENTISTS:** ALAN GUTH

By the late 1970s, space exploration had made it apparent that the Big Bang theory could not explain several observed features of the cosmos. Among the mysteries was the 'flatness problem', which questioned why space was expanding so evenly in all directions. The Big Bang theory predicted that space should not be so uniform unless it had expanded and cooled faster at the start of its existence.

The answer came in the shape of the theory of cosmic inflation, as proposed in 1980 by American theoretical physicist Alan Guth (1947–). In his hypothesis, during the first $10^{-35}$ of a second of time, the Universe doubled in size 100 times over, expanding faster than the speed of light. Thus the Universe would have begun at a billionth of the size of a proton – small enough for the content of the whole Universe to be evenly mixed throughout – and ended up the size of a marble, from which it has expanded more slowly to its current size.

**KEY DEVELOPMENTS**

**Guth's theory of inflation is very well developed and answers all the inconsistencies thrown up by the Big Bang theory. More detailed surveys of the sky since the early 1980s, particularly of the cosmic microwave background, have found that the Universe is indeed extremely homogenous, as predicted by Guth. Although there is no directly observed evidence for the theory, it is still the best explanation we have.**

The evolution of the Universe, starting with the Big Bang. The red arrow marks the flow of time. Note the consequent expansion.

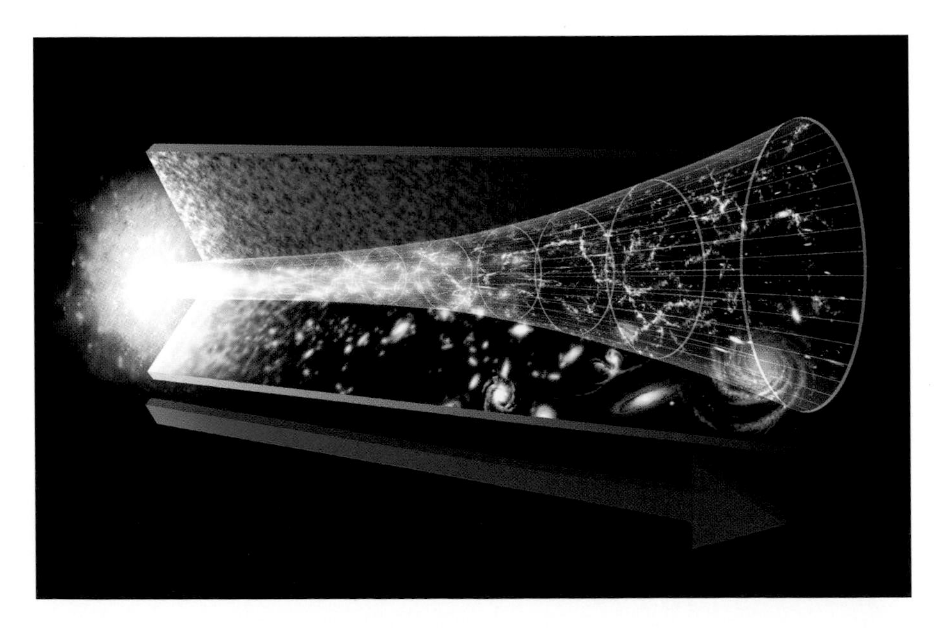

THE EXPANDING UNIVERSE **p.132** COSMIC MICROWAVE BACKGROUND **p.146**

# The Many Worlds Interpretation

**KEY SCIENTISTS:** HUGH EVERETT

An abstract illustration that imagines the Big Bang and the birth of multiple Universes.

Quantum theory is famously weird: events at the quantum level are at the whim of chance. This is the gist of the early twentieth-century Copenhagen interpretation, which states that until a quantum object is observed, and its characteristics measured, it can exist in any number of states. Physicists had previously represented the object as a wave function, which is a set of hypotheticals, each with an attached probability of being the one observed. This wave function 'collapses' when the object is reduced to a firm set of observed properties, and until then all its possible states are in superposition, meaning, for example, that a particle can be in several places all at once.

In 1957, American physicist Hugh Everett (1930–1982) rebelled against this interpretation. He believed that wave function was not a hypothesis but a real thing, and ventured that the whole Universe is an unimaginably complex single wave function.

**KEY DEVELOPMENTS**

**Central to Everett's interpretation is the idea that the wave function does not collapse. It merely seems to disappear because the reality in this Universe becomes decohered with the wave function of its sibling Universes. This decoherence occurs because differences between local realities begin to alter the local wave functions in such a way that they no longer match those of neighbouring worlds or Universes.**

DOUBLE SLIT EXPERIMENT **p.86**

# Anthropogenic Climate Change

**KEY SCIENTISTS:** EUNICE NEWTON FOOTE • SVANTE ARRHENIUS • CHARLES KEELING

American scientist, inventor and women's rights campaigner Eunice Newton Foote carried out early research into the phenomenon now known as the greenhouse effect.

Since 1880, Earth's average temperature has increased by 0.8 degrees Celsius. On a global scale this tiny rise represents a vast boost to the energy driving climate change, and further increases will cause ever-more extreme weather.

The mechanism by which global warming occurs is the entirely natural greenhouse effect. Earth's atmosphere is largely transparent to sunlight, which is absorbed by the planet's surface and radiated away again as heat. However, heat cannot simply shine out into space; greenhouse gases, such as carbon dioxide, trap it in atmosphere, keeping Earth's average temperature at 14 degrees Celsius. Without greenhouse gases, Earth would be a mostly frozen place. Carbon dioxide is released into the air by all life forms, while plants take it out again as a raw ingredient for making sugar by photosynthesis. This 'carbon cycle' maintains the carbon in the atmosphere at a near-constant level.

Fossil fuels represent carbon that was removed from the carbon cycle in the distant past. Burning it for industrial purposes has seen atmospheric carbon increase by a third in 250 years, and there is a well-documented link between this rapid rise in gas and the observed global heating.

**KEY DEVELOPMENTS**

**The link between carbon dioxide and the temperature of Earth arose from the overlooked work of Eunice Newton Foote (1819–1888), who showed that pure carbon dioxide warmed up more rapidly in sunlight than other gases. Evidence of human-caused global heating was provided in the 1960s by her fellow American Charles Keeling (1928–2005). That part of the scientific debate is really over. The focus now is on modelling future climates to promote preventative measures.**

ENVIRONMENTAL SCIENCES p.35

# Origin of the Solar System

**KEY SCIENTISTS:** RODNEY GOMES • HAL LEVISON • ALESSANDRO MORBIDELLI KLEOMENIS TSIGANIS

The inner Solar System contains four terrestrial planets of rock and metal, the largest of which is Earth. The outer four planets are giant worlds made from gas and ice. Additionally, there are millions of smaller rocks in the Asteroid belt, and millions more comet-like ice bodies beyond Neptune in the Kuiper belt, which includes Pluto. It is assumed that this arrangement arose from the way the Solar System formed from a disc of dust, ice and gas left over from the formation of the Sun. The denser materials, such as rock and metals that remain solid at higher temperatures, sank into orbits close to the Sun. The more volatile, low-density substances drifted further out to cooler areas and formed the ice-cold giant planets. As material collided, it coalesced into larger and larger bodies known as planetesimals. Gravity pulled these growing objects into spherical shapes, and the largest bodies got steadily bigger by sweeping up the smaller ones. Once they had cleared their orbits of all other bodies, they became the eight planets. The Asteroid and Kuiper belts represent the leftover rock and ice respectively.

**KEY DEVELOPMENTS**

When astronomers began to observe the Solar Systems of other stars in the late 1990s, they saw something odd. It appeared that gas giants were often orbiting very close to their stars. These so-called 'hot Jupiters' caused a rethink on the formation of our own system called the Nice model (for the French city). This contains a mechanism by which the giant planets began much nearer the Sun, and every time they pulled a planetesimal in from further out, the gravitational reaction made them edge slowly out towards their current orbits.

An artist's impression of the formation of a protoplanetary disc, the precursor of our – and perhaps any – Solar System.

THE SPACE RACE p.32 EXOPLANETS p.148

# METHODS AND EQUIPMENT

# Scientific Process

**KEY SCIENTISTS:** WILLIAM OF OCKHAM • FRANCIS BACON • KARL POPPER • THOMAS KUHN

**KEY DEVELOPMENTS**
**Bacon's process was the encapsulation of ideas that had been evolving since the days of Aristotle. Ockham's razor, which states that the simplest explanation is probably the correct one, dates from the Middle Ages and is still a useful tool today. Philosopher Karl Popper (1902–1994) updated the process in the 1930s by clarifying that science does not reveal a truth as such, but an idea that has not yet been proven to be false.**

A universal man, Francis Bacon was a philosopher, a statesman, a scientist and an author.

Science is a powerful tool for investigating nature and revealing new knowledge. It is performed in five steps. First, scientists observe nature, searching for a question that has no known answer. They then research what is already known about the subject. In the third step, it is time to propose an explanation, or hypothesis, for the problem. Step four sees scientists devise an experiment to test that hypothesis. To do that they must use their theory to predict the results of the experiment. The final step is to interpret the results and draw conclusions about the truth or falsehood of the hypothesis. No investigation carried out like this is a failure; even a negative result will reveal something meaningful. The scientific process is generally credited to the English scholar Francis Bacon (1561–1626), who set out a 'new instrument of science' in 1620. This inspired others, including Robert Boyle, Edmond Halley and Robert Hooke, who would be key figures in the Scientific Revolution that followed later in the seventeenth century.

ALCHEMY **p.15** ISLAMIC SCIENCE **p.16** THE SCIENTIFIC REVOLUTION **p.18**
SCIENCE AND THE PUBLIC GOOD **p.29**

# Graphs and Coordinates

**KEY SCIENTISTS:** RENÉ DESCARTES • ISAAC NEWTON • GOTTFRIED LEIBNIZ

Being able to represent the results of scientific experiments and observations in a graphical way is a great help to analysing their meaning and presenting conclusions. Additionally, by plotting each result as a point on the graph, the data can be converted into a line or another geometrical object. This way of working, known as synthetic geometry, was pioneered by René Descartes (1596–1650). As a result, the basic x–y coordinates used in simple graphs are described as 'Cartesian' in honour of the Frenchman. Descartes is said to have been inspired to create this system while lying in bed, watching a fly trace a route over the ceiling with regular pauses. The coordinates of the pausing places created a mathematical description of the insect's path. Synthetic geometry does the same thing: by converting data into a line, scientists can use algebraic techniques to interpret the information set out. For example, the steepness of a line indicates the rate at which recorded values are changing, while the frequency of any oscillations is revealed by the rising and falling of a line.

**KEY DEVELOPMENTS**

**A generation after Descartes, Isaac Newton and Gottfried Leibniz developed calculus, a way of analyzing data that were in constant flux, as is most often the case with natural phenomena. The process was developed independently, and fiercely defended, by both parties, but both worked by freezing the data at one infinitesimally small point on the graph to show the rate of change at that precise time.**

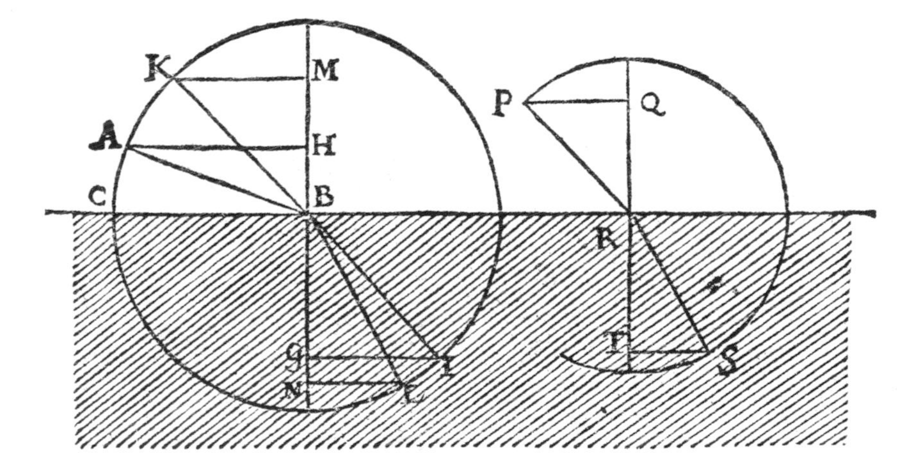


This diagram is taken from the first edition of *Discourse on Method* (1637), in which Descartes used mathematics to formulate a method of deductive reasoning that is applicable to all sciences.

THE SCIENTIFIC REVOLUTION **p.18** PENDULUM LAW **p.52** ACCELERATION UNDER GRAVITY **p.55**
UNCERTAINTY PRINCIPLE **p.166**

# Probability Theory and Uncertainty

**KEY SCIENTISTS:** PIERRE DE FERMAT • BLAISE PASCAL

No measurement of a natural phenomenon, be it the body temperature of a mouse or the energy output of a star, is ever accurate. Therefore, scientists must use statistical tests to show whether the data collected are significant or merely a random collection of unconnected measurements. At the heart of statistics is probability theory – a field of applied mathematics that ascribes values to future outcomes based on the likelihood that they will happen. This is calculated as a proportion of all the possible outcomes.

Probability theory took shape in 1654 when a gentleman gambler presented a question about betting strategies to Blaise Pascal (1623–1662) and Pierre de Fermat (1607–1665), two of Europe's leading thinkers. Probability theory reveals that our natural intuitions about chance are often incorrect. For example, in a coin toss, the chance of getting a head does not increase as the number of tails thrown increases; the chance is always constant. Additionally, the theory shows that in variable phenomena, such as the heights of a group of people, the average values are more commonly represented than extreme ones (short and tall). This spread of data is termed 'the normal distribution', or sometimes 'the bell curve', and it is a valuable tool in assessing whether an observation conforms to the average or differs from it in a significant way.

**KEY DEVELOPMENTS**

**By the 1920s, investigations into the nature of subatomic physics revealed that nature is probabilistic. This means that a particle has the potential to be in a number of states at any one time, with some states being more likely than others.**

French mathematician Pierre de Fermat made many discoveries about the properties of numbers, probabilities and geometry.

PUBLIC HEALTH **p.26** THE NEW PHYSICS **p.27** WAVE-PARTICLE DUALITY **p.128**
LAWS OF THERMODYNAMICS **p.160** QUANTUM PHYSICS **p.167**

# Standard Measurements

**KEY SCIENTISTS:** JOSEPH-LOUIS LAGRANGE • PIERRE-SIMON LAPLACE

The platinum-iridium cylinder on the right was for many years the standard US kilogram. On the left, beneath the bell jar, is its French equivalent.

It is a crucial part of the scientific process that experimental observations can be compared and replicated, and to do that scientists use a system of measurement known as SI units (short for Système International).

There are seven SI base units from which all other measurements, such as speed and force, are derived. These are for time (second; s); distance (metre; m); mass (kilogram; kg); electric current (ampere; A); temperature (kelvin; K); amount (mole; mol, which is usually used to count atoms); and luminous intensity or power of a light source (candela; cd).

The SI system has recently superseded some of its older definitions and now calibrates such units against unchanging physical phenomena. A second of time is based on the oscillation rate of an atom of caesium-133; a metre is the distance travelled by a photon every 3.335641 nanoseconds. The kilogram is based on the amount of energy stored in matter, which in turn is calculated using time and distance. The ampere and candela are similarly defined by other physical constants based on the behaviour of energy at the atomic level. One mole is $6.02214076 \times 10^{23}$, the number of hydrogen atoms in one gram of the gas.

**KEY DEVELOPMENTS**

**An early example of a standardized measurement was the Egyptian cubit, which was a length unit based on the pharaoh's forearm. The US system of measurements evolved by way of British Imperial units from the Roman system, where units were generally divided into twelfths, or an uncia. Both the words 'inch' and 'ounce' come from this division.**

THE RISE OF THE SCIENTIFIC INSTITUTION **p.19** THE SIZE OF EARTH **p.44** WEIGHT OF THE EARTH **p.74**

# Measuring Time

**KEY SCIENTISTS:** CHRISTIAAN HUYGENS • JOHN HARRISON

**KEY DEVELOPMENTS**

Everyday measurements of time are based on the motion of the Sun, using both the day–night cycle and the year. The ancient Egyptians divided daylight into 10 hours, and added one more hour for dawn and another for dusk, thus creating the 24-hour day and night. A month was initially based on the time between one full moon and the next, roughly divided into four weeks to match the progression through the four lunar phases.

The science of time measurement is named horology. While many fundamental questions about the nature of time – even if it really exists – are among the enduring mysteries of physics, horologists have developed increasingly accurate timepieces to record its passing. Accurate clocks are always based on a physical oscillation with a constant period (the time taken to complete each oscillation). An early example was the swing of a pendulum, but today clocks are more likely to use a crystal of quartz. Quartz is a piezoelectric substance, which means that it vibrates when electrified, and that rhythmic distortion is used to control a pulse of electricity that counts out a known tiny fraction of a second.

Specialist atomic clocks are even more accurate: in them, the oscillation is a change in the way the atoms of caesium (or atoms of a similar metal) absorb and emit energy. While a good quartz timer might add or lose just a few seconds every year, an atomic clock will only become a second out after 100 million years (at least).

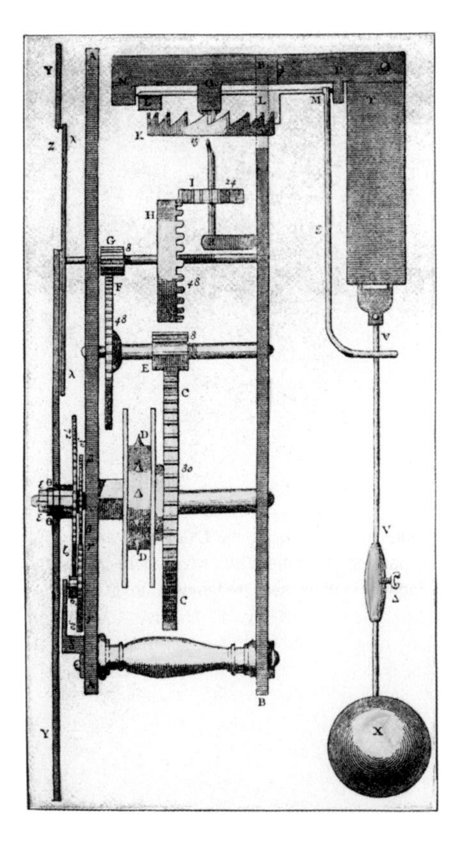

This hand-coloured woodcut shows a side view of the pendulum mechanism of one of the earliest clocks produced in the 1660s by Christiaan Huygens.

PENDULUM LAW **p.52** SPEED OF LIGHT **p.98** ROTATION OF EARTH **p.100** RELATIVITY **p.163**

# Thermometers

**KEY SCIENTISTS:** GALILEO GALILEI • OLE RØMER • DANIEL GABRIEL FAHRENHEIT ANDERS CELSIUS • LORD KELVIN

A late nineteenth-century medical thermometer calibrated in Fahrenheit, the scale named after Daniel Gabriel Fahrenheit (1686–1736), the Polish Dutch scientist who invented it, in which water freezes at 32 degrees and boils at 212 degrees.

Heat, or thermal energy, is the manifestation of atoms and molecules moving. The faster they move around (or wobble back and forth), the more energy they contain. Temperature is a measure of the average thermal energy in a substance. A thermometer measures temperature using a scale, which has entirely arbitrary upper and lower points. The most successful scales base these points on temperatures that are easy to replicate. For example, the Celsius scale sets these points at the freezing and boiling points of water. One hundred equal units, or degrees, denote the temperatures in between. A digital thermometer uses a thermistor – an electronic component that changes its resistance as the temperature varies. The laser in a laser thermometer is just an aiming aid; the device is measuring the infrared waves emitted from a body or other object. Earlier thermometers, now largely defunct, used a liquid, such as mercury or alcohol, that would respond to temperature changes by expanding and contracting inside a narrow calibrated tube.

**KEY DEVELOPMENTS**

**The first devices for measuring the 'temper of the air' used sealed columns of water. Water was replaced by mercury in the 1700s because the expansion of the metal was much more consistent. Today scientists measure temperature in Kelvin (K). An increase of 1K is of the same magnitude as an increase of one degree Celsius. However, zero degrees Kelvin (0K) is set at the lowest possible thermal energy of subatomic particles, which is absolute zero.**

GAS LAWS **p.60** MECHANICAL EQUIVALENT OF HEAT **p.96** LAWS OF THERMODYNAMICS **p.160**

# Microscopes

**KEY SCIENTISTS:** ROBERT HOOKE • ANTONIE VAN LEEUWENHOEK

The idea of using a lens to magnify small details dates back to at least 700 BCE, when transparent crystals were carved into the characteristic curve shape. The smooth curve of a lens means that rays of light all arrive at slightly different angles, and thus are refracted (have their direction altered) by correspondingly varying amounts. The result is that all the rays are focused onto a single point on the other side of the lens. Holding the lens to the eye so that the focal point coincides with the retina creates the illusion of looking at a much larger 'virtual image' of the original – revealing its hidden details. A microscope uses at least two lenses to boost this effect. The lower, objective lens makes a sharp and clear 'actual' image of the object within the microscope. The eyepiece lens then magnifies this into a much larger virtual image, potentially many hundreds of times larger than the original object. Optical microscopes are equipped to illuminate specimens from above or below, and use filters to apply different qualities of light to reveal features. Additionally, biological specimens are often dyed with chemicals that highlight specific regions of the tissue or cell.

**KEY DEVELOPMENTS**

**Light microscopes are limited to a maximum magnification of about × 2000. Above that, small objects appear fuzzy and hard to tell apart when closely packed. In the 1930s, the electron microscope (EM) was developed to greatly boost this limit. Images created using beams of electrons can resolve objects many thousands of times smaller. The scanning transmission EM can detect objects, including individual atoms, just 50 billionths of a metre wide.**

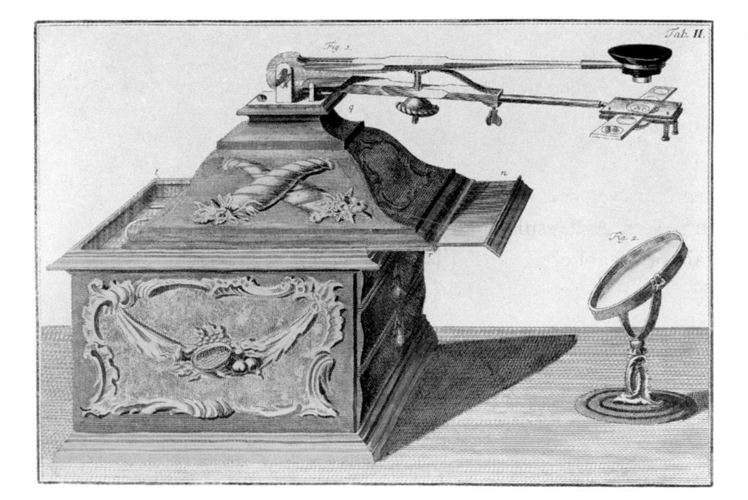


A microscope of 1763 built by German biologist Wilhelm Friedrich von Gleichen-Russwurm (1717–1783).

CELL THEORY **p.25** GENETICS **p.31** DISCOVERY OF MICROORGANISMS **p.64**

# Telescopes

**KEY SCIENTISTS:** HANS LIPPERSHEY • GALILEO GALILEI

FAST, the world's largest radio telescope, in Guizhou Province, China, is 500 metres (1640 feet) in diameter.

The utility of telescopes was proved by seventeenth-century astronomers who were able to see many more smaller and distant objects and thus began to grasp the extent of the Universe. Those early devices were refracting telescopes. Unlike in a microscope, the larger lens is at the front. This objective lens is as big as possible to collect as much light from as far afield as it can. It focuses this light into a clear image within the telescope's barrel, and the eyepiece then magnifies this so it can be seen. As a result, a telescope reveals details of distant objects, such as planets, and shows up other objects that are too dim to see with the naked eye. Binoculars are essentially two refracting telescopes side by side. Reflecting telescopes replace the objective lens with a curved mirror to collect light. This design is easier to engineer on a large scale, and so is used for the most powerful telescopes, such as the European Extremely Large Telescope and the Hubble Space Telescope.

**KEY DEVELOPMENTS**
**It is possible to image the Universe using radiation. A radio telescope works like a large antenna collecting signals from space, and reveals interesting deep-space objects such as quasars. X-ray and ultraviolet telescopes are best placed in space because their rays are blocked by the atmosphere. Once launched, the James Webb Space Telescope will view the Universe by its heat signature and see further than ever before.**

CAMERA OBSCURA **p.46** HERTZSPRUNG-RUSSELL DIAGRAM **p.122** COSMIC MICROWAVE BACKGROUND **p.146**
EXOPLANETS **p.148** LIGO **p.152** DARK MATTER **p.175**

# Microphones and Speakers

**KEY SCIENTISTS:** THOMAS EDISON • ALEXANDER GRAHAM BELL

**KEY DEVELOPMENTS**

Acoustics – the science of sound – seeks to investigate the way pressure waves move through gases, liquids and solids, each of which have a specific speed of sound largely dependent on their densities. The earliest discovery in acoustics is credited to Pythagoras, the ancient Greek mathematician. He noted that the rising musical tones of an octave involve dividing the first note's wavelength by two, then by three, four, five, six and seven.

The technology to record and play back sounds arose in the late 1800s as an arm of the burgeoning entertainment and communications industry. It has since become a ubiquitous tool in science as a means of collecting and analysing information about all kinds of natural phenomena, from animal communication to mapping the sea-floor with sonar.

Microphones convert sound-wave oscillations in the air (or other medium) into corresponding electrical signals that can be transmitted, amplified and recorded. In the simplest design of microphone, pressure waves of sound hit a diaphragm, making it wobble with the same rhythm. This wobble moves magnets relative to one another, and that induces a fluctuating electric current that serves as the sound signal. The loudspeaker takes that signal and uses it to electrify an electromagnet, which then attracts and repels another magnet to create a rhythmic motion. This movement is transferred to a cone which wobbles the air, creating a sound wave with the same structure as the original.

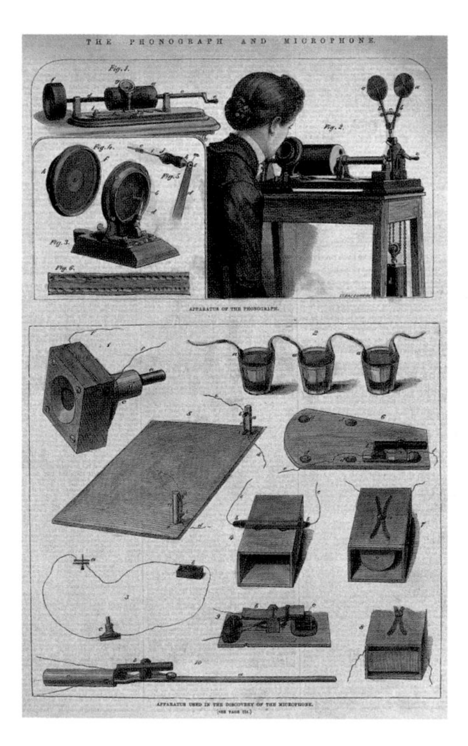


This coloured wood engraving shows an Edison phonograph with a carbon microphone.

GREEK PHILOSOPHERS **p.13** SCIENCE AND THE PUBLIC GOOD **p.29** ELECTROMAGNETIC UNIFICATION **p.88**

# Geiger-Müller Tube

**KEY SCIENTISTS:** HANS GEIGER • WALTHER MÜLLER

Fully named for Hans Geiger (1882–1945) and Walther Müller (1905–1979), who invented it in 1928, this device is more commonly known as a Geiger counter. Geiger developed its working principle while he was involved in the discovery of the atomic nucleus. The device detects ionizing radiation. This is generally the fast-moving particles and rays produced by radioactive atoms that have enough energy to knock electrons off atoms in the air, creating charged ions. The high energy of such emissions makes them dangerous to health. The tube is filled with a diffuse gas at about one tenth of atmospheric pressure. In normal conditions, two electrodes inside the tube are unable to send an electric current through the gas. However, when ionizing radiation is present, it creates charged ions inside the tube, allowing a pulse of current to flow. This pulse actuates a loudspeaker, which makes a clicking sound. As the number of ionizations increases, the number of clicks rises accordingly, and those sounds merge to make a tone. The higher the pitch, the greater the danger.

**KEY DEVELOPMENTS**

**While the count of ionizations is a very useful tool, the basic Geiger counter does not indicate how much energy is carried by the radiation. Alpha particles carry much more energy than gamma rays or X-rays, for example. Instead, a dosimeter must be used to measure how much energy a person is exposed to from dangerous radiation.**

This amplifier, linked to a Geiger counter, was used in radioactivity experiments by Frédéric Joliot-Curie and his wife Irène Joliot-Curie, the son-in-law and daughter of Pierre and Marie Curie.

SCIENCE AND THE PUBLIC GOOD **p.29** ENVIRONMENTAL SCIENCES **p.35** DISCOVERY OF RADIOACTIVITY **p.112** NUCLEAR FISSION **p.136**

# Photography

**KEY SCIENTISTS:** NICÉPHORE NIÉPCE • LOUIS DAGUERRE • WILLIAM HENRY FOX TALBOT

The English word 'camera' derives from the Latin for 'chamber', and the earliest cameras were just that: dark rooms that mimicked the workings of the human eye, letting light through a small aperture in one wall to create images of the outside world on the opposite wall. In the nineteenth century, several systems were developed to capture these images using light-sensitive chemicals such as silver iodide.

Until the 1990s, photography was based on plastic film. This technology transformed science, replacing the need to make drawings of observations, and enabling people to make visual records of occurrences that were too quick or too faint for the naked eye to detect.

Digital photography has now largely replaced film, making images using light-sensitive electronics to build patterns of pixels that are recorded as computer data. Digital cameras are much smaller and more durable than their predecessors, and this has led to a proliferation of their use. Moreover, live images can now be sent as data from the camera from remote locations, such as space or the ocean.

**KEY DEVELOPMENTS**

**The persistence of vision is the phenomenon in which the human brain retains an image and creates a seamless moving image out of still pictures in rapid succession. This was demonstrated most successfully in 1887, when photographs of a galloping horse revealed all the animal's feet left the ground as it ran. Photographic technology was soon being used to create movies and video.**

This photograph was made in 1839 by William Henry Fox Talbot (1800–1877). The image was created without a camera using a piece of translucent seaweed laid directly onto a sheet of photosensitized paper, blocking sunlight from the portions it covered and leaving a light impression of its form.

ELECTRONICS AND COMPUTATION **p.30** CAMERA OBSCURA **p.46**

# Cathode-Ray Tube

**KEY SCIENTISTS:** HEINRICH GEISSLER • WILLIAM CROOKES

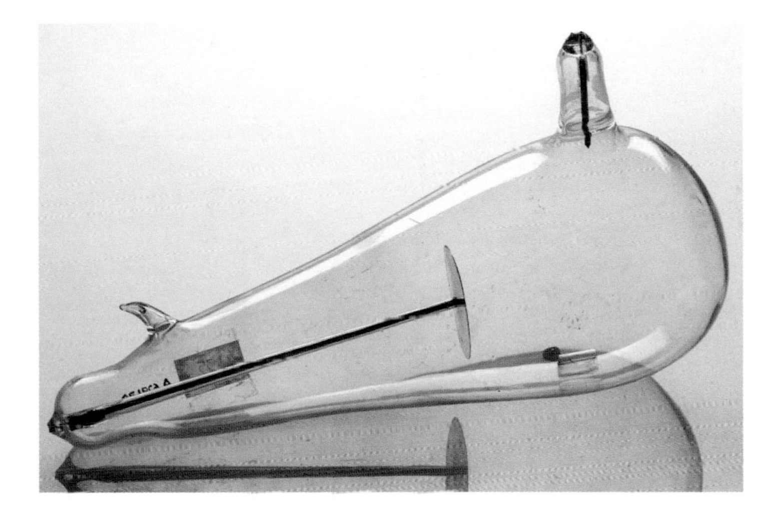

In the 1870s, William Crookes (1832–1919) studied the effects of passing electricity through partially evacuated tubes like this one. Rays emitted from the cathode, which is an aluminium disc in this tube. These cathode rays were later shown to comprise electrons.

A cathode-ray tube (CRT) is a sealed glass tube with a near-vacuum inside and a cathode (a negative electrode) and an anode (the positive one) within it. Largely superseded today by electronic and digital technology, the CRT was a very useful tool in probing the nature of the atom and electromagnetism, not least its role in the discovery of X-rays and electrons.

The CRT evolved from more primitive devices that contained more gas. These glowed with distinctive colours and were early versions of today's fluorescent light bulbs. When the gas was mostly sucked out, a mysterious ray formed seemingly from nothing. This was the cathode ray, which was later shown to be a beam of electrons. Adding electromagnetic coils around the tube made it possible to control the direction of the beam. This innovation created an oscilloscope where the beam produced a dot on a screen that moved around according to the activity of the electromagnets. At their simplest, oscilloscopes create a visual representation of changes in an electric current. However, they can be repurposed to show (and measure) heartbeats, earthquakes, sound waves and all kinds of constantly changing phenomena.

**KEY DEVELOPMENTS**

**The cathode-ray tube was at the heart of the original electronic television set. It created a picture by sweeping the electron beam in lines across a fluorescent screen, making it glow. A flickering beam created an image as a pattern of light and dark spots. The image was renewed and updated many times a second, much faster than the human eye could detect, thus creating the illusion of a moving picture.**

DISCOVERY OF THE ELECTRON **p.114** WAVE-PARTICLE DUALITY **p.128** ATOMIC THEORY **p.159**

# X-Ray Imaging

**KEY SCIENTISTS:** JAMES CLERK MAXWELL • HEINRICH HERTZ • WILHELM RÖNTGEN

**KEY DEVELOPMENTS**
X-ray machines were the first medical scanners to offer doctors a view of the internal anatomy of their patients. They are widely used today, although the X-ray images only show up hard body parts, or those treated with some kind of contrasting agent. Computed tomography (CT) scans use several X-ray images taken at varying angles to create a more detailed image. X-rays have enough energy to damage cells, and frequent exposure is dangerous. However, occasional X-ray scans carry a negligible risk.

In the late 1880s, physicists were investigating the properties of cathode rays. These were eerie glowing beams created by electrifying a vacuum tube. The beams had the same effect on photographic paper as ordinary light: they made it darken and appear foggy. In 1895, German physicist Wilhelm Röntgen (1845–1923) found that his electrified tube was fogging photo paper without giving out the expected glow. Obviously this was an invisible ray of some kind, and Röntgen recorded his mysterious discovery as X-rays – the name stuck. Investigating further, Röntgen found that X-rays could pass through some solids, and he took an X-ray photograph of his wife's hand, revealing the bones inside in the now familiar way. X-rays have much higher energy than visible light, and so they penetrate through skin and soft tissue but are blocked by hard bone. Therefore, the bones cast a shadow on the paper, leaving it white, while the X-rays fog the rest. The following year, the first X-ray imager was set up at a hospital in Glasgow, Scotland, and by 1898 the British Army were using a mobile X-ray unit in Sudan to scan injured soldiers behind the battle lines.

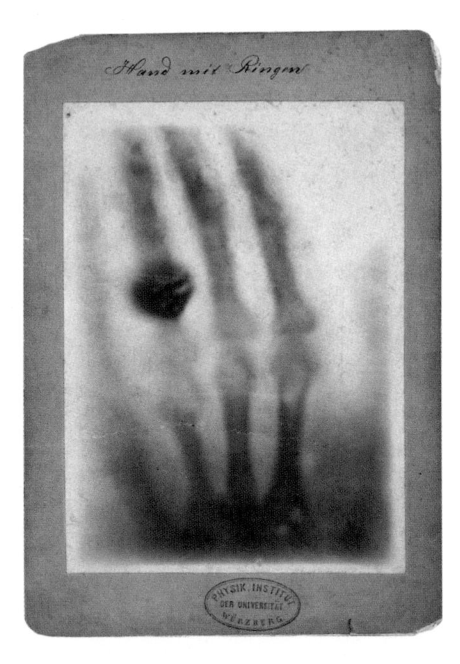


When Anna Röntgen saw the image that her husband had made of her hand (and wedding ring) she was unhappy with the results, saying, 'I have seen my own death'.

THE NEW PHYSICS **p.27** SCIENCE AND THE PUBLIC GOOD **p.29**
DISCOVERY OF ELECTROMAGNETIC WAVES **p.110**

# Lasers

**KEY SCIENTISTS:** CHARLES TOWNES • GORDON GOULD

Natural sources of light, such as those from the Sun or flames, are a mixture of colours, or wavelengths, and the beams are incoherent – their light waves are vibrating out of sync. A laser is a source of light that contains just one or two wavelengths, and the beam is coherent; its waves are all precisely synchronized. A coherent beam can be manipulated (for example, reflected or divided) in very precise ways, and hence has many uses. The word laser is an acronym of 'light amplification by stimulated emission of radiation'. The first version was actually a maser, a microwave, but the light laser was invented in 1960. It works by injecting light into a crystal – originally a ruby, but today most laser crystals are synthetic substances. Certain photons in the light are absorbed by the crystal's atoms, which then re-emit them and excite more and more atoms. The crystal is surrounded by a mirrored surface, and so the light inside bounces around, building in intensity. This light is then released from one end of the device as a powerful laser beam.

**KEY DEVELOPMENTS**

Lasers have numerous uses. They can read and send data, as when they reflect from a barcode or a spinning DVD or flicker out a coded signal along an optical fibre. Lasers are also used as precise measurement tools. They were regularly bounced off mirrors placed on the Moon by Apollo crews. The time of the laser beam's flight there and back, all at the speed of light, gave a very accurate distance between Earth and the lunar surface. Lasers can also deliver energy for use in surgery in place of scalpels, in industry in place of drills and saws, and even as weaponry.

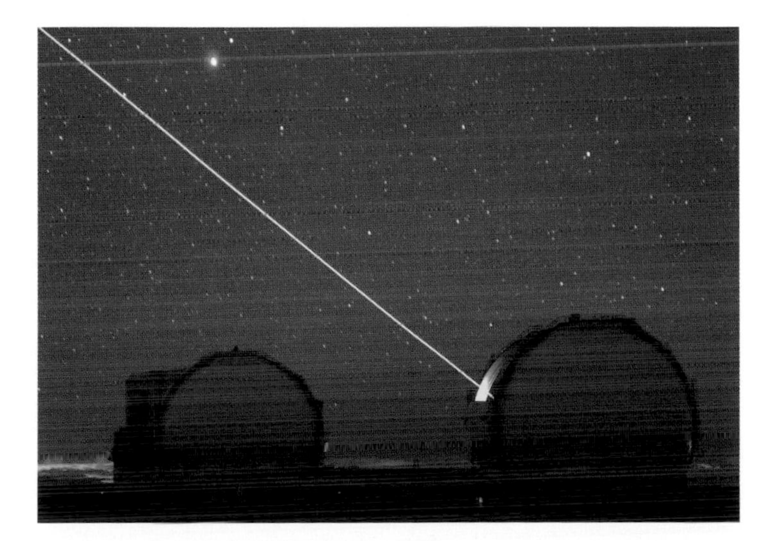

This laser beam from the Keck 2 telescope in Hawaii, USA, is used to measure the effects of Earth's atmosphere in making starlight appear to shimmer and twinkle.

THE INTERNET **p.36** LIGO **p.152**

# Seismometers

**KEY SCIENTISTS:** ZHANG HENG • JEAN DE HAUTEFEUILLE • ALFRED EWING CHARLES RICHTER

A seismograph machine needle drawing a red line on graph paper depicting seismic and earthquake activity.

A seismometer is a device that measures vibrations travelling through the ground. These vibrations are caused by rocks cracking underground and are particularly associated with earthquakes. The first seismometer was built in 132 CE in China. It had a central pendulum which, once set swinging by a tremor, released one of several balls positioned around the device, indicating the direction of the quake. Similar mechanical seismometers were refined over the centuries, tracing the vibrations on paper. Today, the same job is performed by sensitive electronics. During the nineteenth and twentieth centuries, seismometers were set up all over the world and were able to pick up vibrations from the major earthquakes that happened every few months somewhere in Earth's crust. The seismic waves produced are strong enough to move right through the planet, and by comparing how different waves were reflected and refracted on their subterranean journey, geologists found that Earth has a dense metallic core and an outer mantle layer made from molten rocky material.

**KEY DEVELOPMENTS**

**The power of an earthquake is popularly reported using the Richter scale, developed by Charles Richter (1900–1985). The scale runs from magnitude 1, which represents more or less continuous microquakes only detected by seismometers, to greater than magnitude 9, which is felt perhaps once every 50 years. The scale is logarithmic, so an increase in one magnitude represents a tenfold increase in power.**

GEOLOGY AND THE EARTH SCIENCES **p.23**

# Radiocarbon Dating

**KEY SCIENTISTS:** WILLARD LIBBY • ARTHUR HOLMES

Natural levels of radioactivity can be used to determine the age of ancient objects. Every radioactive isotope has a specific half-life, which is the time taken for half its atoms to decay away. This decay is infinitely divisible: if a certain isotope has a half-life of one week, it will lose half its atoms in the first seven days, and then half its remaining atoms in the second seven days and so on ad infinitum.

As an object gets older, the proportion of these isotopes reduces, and determining by how much will give a good estimate of the object's age. Organic objects – bodies, tools made of wood and bone, fabrics woven from natural materials – can be dated using carbon, an element found in all living things. Carbon-14, a radioactive form of carbon, exists naturally in small amounts. During life, the proportion of this isotope stays the same in the body, being constantly renewed. But after death, the carbon-14 level drops, as the atoms break down into a stable form of nitrogen. With a half-life of 5,730 years, carbon-14 can be used to date objects that are up to 50,000 years old.

**KEY DEVELOPMENTS**

Rocks and minerals, which can be millions if not billions of years old, are dated using uranium and other radioactive metals with much longer half-lives than carbon-14. This form of radiometric dating, first carried out in the early twentieth century, provided the first proof of the long-held intuition that Earth was far more ancient than could be determined using human historical records.

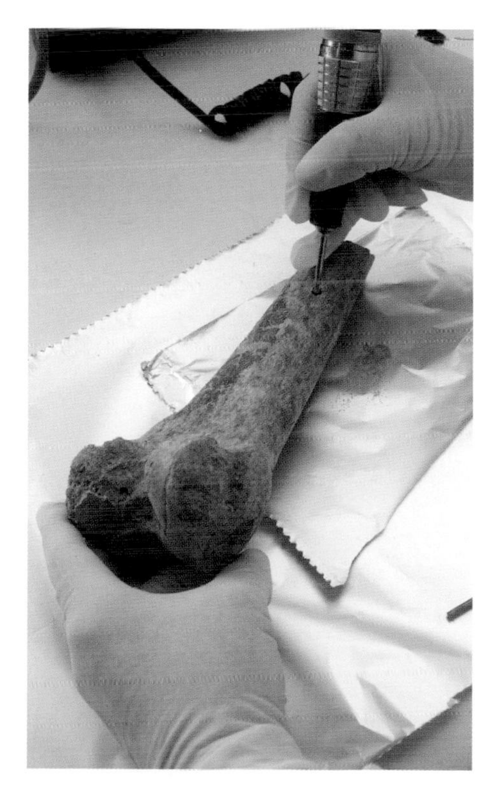

Removing organic material from the interior surface of an ancient human bone to use as material for radiocarbon dating.

NATURAL HISTORY AND BIOLOGY **p.22** GEOLOGY AND THE EARTH SCIENCES **p.23**
DISCOVERY OF RADIOACTIVITY **p.112**

# Bubble Chambers

**KEY SCIENTISTS:** DONALD A. GLASER • CHARLES WILSON

**KEY DEVELOPMENTS**

In 1928, British quantum physicist Paul Dirac (1902–1984) developed an equation that described the full behaviour of electrons. However, the mathematics revealed that it would be possible for another particle to exist with the same mass but otherwise wholly opposite properties. This was named the positron, a positive version of the electron, and it was the first example of antimatter. Antimatter is very short-lived. When it meets matter, it is annihilated into nothingness.

The discovery of subatomic particles at the start of the twentieth century led to a technological problem: how to observe these almost infinitesimal entities. The problem was compounded by the discovery, from observations of cosmic rays hitting the atmosphere, that most subatomic particles are very short-lived, visible for only a split second after high-energy collisions. In 1911, the cloud chamber was invented in which particles whizzed through thick water vapour. The particles electrified the molecules, creating a trail that was captured in a photograph. By the 1950s, this system had been updated to a bubble chamber, inspired, legend has it, by the way bubbles form in beer. The chamber holds super-chilled liquid hydrogen at a high pressure and just at its boiling point. Passing particles make the hydrogen boil, leaving the telltale trail. Electric and magnetic fields are applied across the chamber, and the precise curved path of a particle indicates its charge, mass and energy. Bubble and cloud chambers were instrumental in the discovery of exotic particles such as positrons and mesons. By the end of the twentieth century, these chambers had been largely superseded by electronic detectors.

An old bubble chamber displayed at the Fermi National Accelerator Laboratory (Fermilab) in Illinois, USA, where scientists research the smallest building blocks of matter.

THE NEW PHYSICS **p.27** DISCOVERY OF THE ELECTRON **p.114** COSMIC RAYS **p.124** THE STANDARD MODEL **p.174**

# Particle Accelerators

**KEY SCIENTISTS:** ERNEST LAWRENCE • M. STANLEY LIVINGSTON • GLENN SEABORG

M. Stanley Livingston (left; 1905–1986) and Ernest O. Lawrence (1901–1958) in front of their creation: the 68-centimetre (27-inch) cyclotron at the old Radiation Laboratory at the University of California, Berkeley (1934).

Physicists have so far found that the Universe is made from 18 fundamental particles, most of which were discovered by smashing matter together in powerful collisions that destabilized the atoms. Scientists then watched as the matter reorganized itself into a stable form, thus providing clues to how the Universe formed.

All this science takes place inside particle accelerators, which have been constantly improved since they were invented in the late 1920s. The heart of the accelerator is a vacuum tube surrounded by powerful magnets, which produce electric and magnetic fields that guide a beam of particles down the tube and give it successive pushes to speed it up. Two beams are then precisely controlled so they smash head-on inside a detector. The largest accelerator today is the Large Hadron Collider (LHC). It can accelerate protons to 299.8 million metres per second, a fraction below the speed of light. At these speeds, adding more energy does not make the protons faster, but heavier. As such, when they collide, the protons in the LHC weigh 7,500 times more than they did when they entered the accelerator.

**KEY DEVELOPMENTS**

**Particle physics is generally carried out using cyclical accelerators like the LHC. However, linear accelerators are used to fire heavier objects – entire atoms – at a target at the far end. These machines are used to explore superheavy elements, where atoms too heavy to exist in nature are created by fusing two smaller atoms together. So far, 25 of these artificial elements have been produced. They are all unstable, but scientists predict that in future heavier atoms created by the same process might be more stable.**

NUCLEAR FISSION **p.136** THE STANDARD MODEL **p.174** DARK MATTER **p.175**

# ATLAS (CERN)

**KEY SCIENTISTS:** PETER JENNI • FABIOLA GIANOTTI • DAVID CHARLTON • KARL JAKOBS

The Large Hadron Collider (LHC) at CERN, a physics research institute on the border of France and Switzerland, has eight detectors. The largest is ATLAS, which was the chief component in the discovery of the Higgs boson in 2012. This new particle, which the LHC was built to discover, gives matter particles, such as electrons and protons, the property of mass. ATLAS weighs 7,000 tonnes and is six storeys high and about as long as a Boeing 737. At its heart is an impact chamber, inside which protons collide with an energy not generated in the Universe since the early moments of the Big Bang. The particles produced by the collision spray out in all directions, each taking a distinctive path as they interact with a strong electromagnetic field that surrounds the chamber. The particles then pass through layers of silicon detectors which map the particles' positions, pinpointing each to one of 80 million possible locations inside ATLAS. Next, the particles are tracked through thicker strips of silicon before moving into outer detectors that can identify the mass, energy and other properties of each particle. ATLAS tracks 40 million proton collisions every second.

**KEY DEVELOPMENTS**

**The LHC is also equipped with the ALICE detector, which investigates what happens when the nuclei of much heavier lead atoms are smashed together. The nuclei contain protons and neutrons which are made of quark particles. Quarks are locked together by gluon particles that control the strongest forces in the Universe. Through the use of ALICE – an acronym of A Large Ion Collider Experiment – scientists can observe how the strong force works to create matter as it occurred in the early stages of the Universe.**

Installing the ATLAS detector at the Large Hadron Collider, CERN, Switzerland.

THE STANDARD MODEL **p.174**

# Neutrino Detectors

**KEY SCIENTISTS:** ENRICO FERMI • WOLFGANG PAULI

During some nuclear reactions, neutrons collapse into protons, releasing electrons. The products from this change do not add up to exactly the same mass as the original particle. Another particle – a neutrino – is also released. Neutrinos have a mass, but it is so tiny that physicists are still trying to measure it accurately, and being so ephemeral, neutrinos hardly interact with other particles. They barely do anything, so their actual properties are hard to study. In an effort to do so, neutrino detectors have been located in places where they are shielded from cosmic rays: on the ocean floor, under Antarctic ice or in deep disused mines. A neutrino passing through such a detector might collide with an atom and release a flash of radiation that is picked up by super-sensitive cameras. Neutrinos are very common, and make up a vast amount of material in the Universe. Each second, about 65 billion of them stream through every square centimetre of Earth's surface. However, even the biggest detectors pick up only about 10 of them per day.

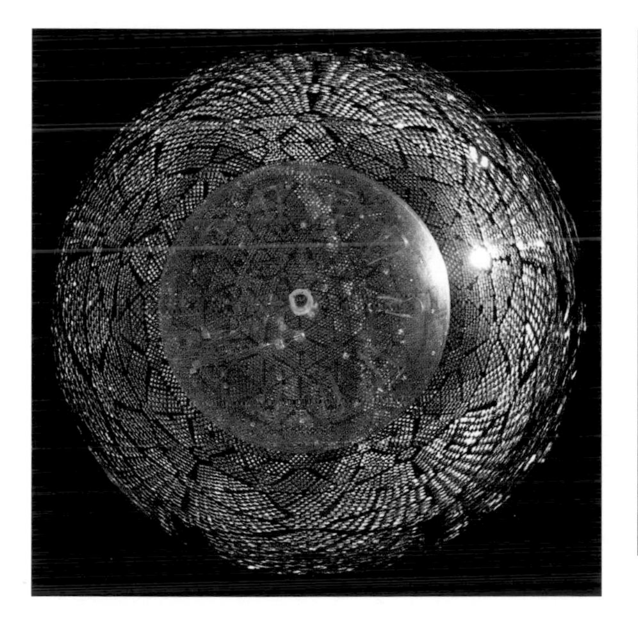

A geodesic sphere installed nearly 2,500 metres (8,000 feet) below the surface of the Creighton mine at the Sudbury Neutrino Observatory in Ontario, Canada.

**KEY DEVELOPMENTS**

As their name suggests, neutrinos have no electrical charge – they are neutral. However, they come in different types or 'flavours'. One type is linked to the electron, and another two are associated with muon and tau particles, which are heavy versions of the electron observed fleetingly in nuclear reactions. To add to the neutrino mystery, it is thought that the particles repeatedly switch from one flavour to another and take on a fourth 'sterile' flavour as they transition. Sterile neutrinos are even harder to detect than the other types.

THE NEW PHYSICS **p.27** THE STANDARD MODEL **p.174** NUCLEAR FISSION **p.136** DARK MATTER **p.175**

# Mass Spectrometry

**KEY SCIENTISTS:** J.J. THOMSON

The mass spectrometer is a device used in physics and chemistry to analyze the composition of a mixed sample of subatomic particles, atoms and molecules. It was developed by J.J. Thomson as an extension of the cathode-ray tube, in which an electric field was used to bend a beam of electrons and thereby measure their weight.

Thomson developed this procedure by introducing a sample of material into the tube and then seeing how its different components were spread out by the electric field in order of their mass. This is the basic principle of a mass spectrometer. A detector at the end of a vacuum tube displays the spread of masses of the objects in the sample and their relative quantities.

Thus Thomson measured the proportions of isotopes – variant versions of an element's atoms, with the same chemical properties but slightly different masses. Chemists now use mass spectrometers to analyze complex chemical mixtures, using the masses revealed as clues to the structure of the molecules present.

**KEY DEVELOPMENTS**

The phenomenon of isotopes was discovered in 1912 by Frederick Soddy (1877–1956), an English physicist. He was working with many other big names in atomic physics to analyze how radioactive atoms like uranium decayed. They found the unstable atoms changed repeatedly through several forms (now understood as a decay series), which they initially saw as new, undiscovered elements. However, Soddy found that they were all already known substances, but they appeared in a variety of slightly different masses, which Soddy named isotopes.

1906 photograph of J.J. Thomson, inventor of the mass spectrometer, discoverer of electrons and isotopes, and pioneer of nuclear physics.

DISCOVERY OF RADIOACTIVITY **p.112** DISCOVERY OF THE ELECTRON **p.114** THE PERIODIC TABLE **p.162**

# Chromatography

**KEY SCIENTISTS:** MIKHAIL TSVET

With a name meaning 'colour writing', this process is familiar to many as a school science experiment, where inks and paints are divided into their constituent colours, thus demonstrating that chromatography is a way of separating a mixture of substances.

The science of mixtures is complex. A heterogeneous mixture, such as sand and seawater, can be separated using a filter because one of the constituents is so much bigger than the other. Seawater, by contrast, is a mixture of salt and water, where the constituents are of similar sizes and so they are mixed homogeneously. Filtering does not work here, but salt and water have distinct physical properties, so the liquid water can be boiled away to leave the pure salt.

Chromatography is used when a homogeneous mixture has substances with very similar physical properties. These substances are made to move through a medium, such as water, gas or gel. This can be done by passive means – for example, by letting the material soak or diffuse naturally – or by applying an electric field. Each of the constituents will move a specific distance, thus separating them out for analysis or collection.

**KEY DEVELOPMENTS**

The isotopes of an element vary from each other by tiny differences in mass, and they are separated and purified using a gas centrifuge, which bears comparison to chromatography. The isotopes are added to the centrifuge as a diffuse gas. The spinning motion flings the atoms outwards, with the lighter isotopes becoming more common near the edge with the heavier ones located further in. This system is used most notably to collect the fissile isotopes of uranium employed in nuclear fuels and weapons.

Paint pigments being separated by simple chromatography.

ALCHEMY **p.15** THE BIRTH OF CHEMISTRY **p.20**

# Distillation

**KEY SCIENTISTS:** JUSTUS VON LIEBIG

Distillation is a technique for separating mixed liquids. It is an ancient technology used most notably to extract pure alcohol from fermented beverages. The process works by controlling the temperature of the mixture so that only the liquid with the lower boiling point evaporates into a vapour, leaving the one with the higher boiling point in its original state.

In pre-industrial times, distillation was performed with a retort, a roughly tear-drop-shaped vessel. The vapour rising from the mixture condensed on the cooler top of the retort and trickled down the long spout for collection. Today, chemistry labs are equipped with condensers in which the vapour is channelled through a tube that is surrounded by a jacket of cold water. The rapid reduction in temperature condenses the vapour inside back into a liquid very efficiently. When distilling mixtures with very similar boiling points, the vapour is channelled into a fractionating column. The lighter materials rise to the top with the heavier ones lower down. Crude oil is separated into its constituents using fractional distillation on an industrial scale.

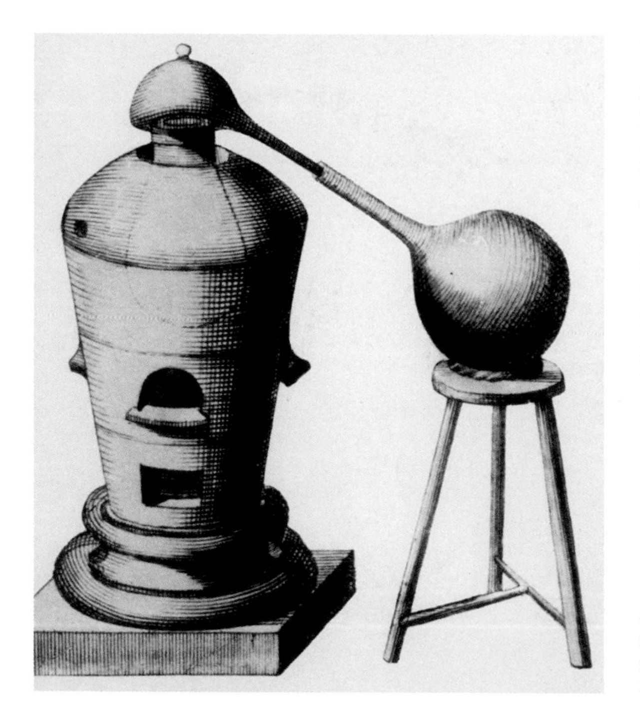

**KEY DEVELOPMENTS**

**The gaseous constituents of air are also separated by distillation, but in this case the method involves refrigeration rather than heating. Once small amounts of carbon dioxide and water have been removed, the first gas to liquefy is nitrogen. Then comes oxygen, leaving about 1 per cent that is composed of valuable inert gases such as argon and neon.**

Alembics such as this eighteenth-century example may be used to separate substances into their constituent parts.

ALCHEMY **p.15** ISLAMIC SCIENCE **p.16** THE BIRTH OF CHEMISTRY **p.20**

# DNA Profiling

**KEY SCIENTISTS:** ALEC JEFFREYS

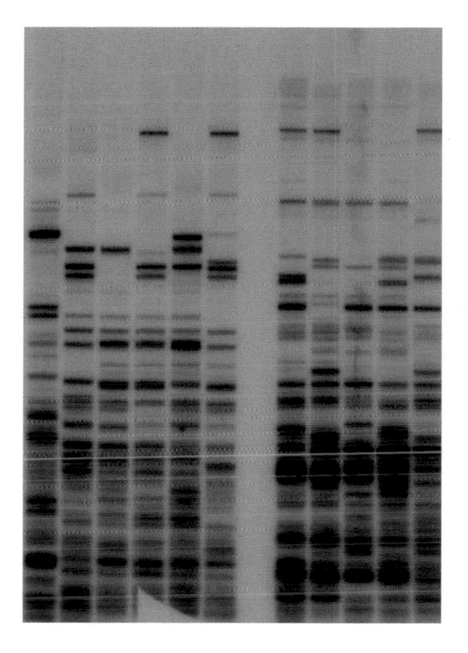

**KEY DEVELOPMENTS**

All humans share 99.9 per cent of their DNA. Nevertheless, every person's DNA carries a unique genetic code, but their DNA profiles are not unique. The chance of two people having the same DNA profile is about one in five million. It is therefore possible that a crime-scene profile matches that of a suspect by coincidence. Juries are directed to consider whether that is likely for a defendant connected to crime in other ways.

An early DNA fingerprint showing DNA patterns from a mother (lanes 2 and 8) and her four children (adjacent lanes to the right). Note the similarities between these lanes and their differences from lane 1, which is the DNA of an unrelated person.

The DNA profile, or genetic fingerprint, revolutionized forensic science when it was invented in 1984 because it could provide highly compelling material evidence that a suspect had been present at a crime scene. The technique has made it faster to confirm disputed parentage. It is also used in medical testing and research, plus in calculating the genetic variability of populations of endangered animals.

Contrary to a popular belief, the profile does not provide a person's genome or a complete decoding of a person's genes. Instead, it builds a highly specific pattern based on particular features in a person's genetic code. The process searches for sections of genetic code that repeat multiple times. Everyone has these repeats, but the length and location of them varies from person to person. The repeated regions are copied many times in the lab, and then separated on a gel and dyed so that they create a barred pattern of varying thicknesses. This pattern can be used to match to another profile to show that the DNA is from the same person, or it can be compared with the profile of a family member to prove a genetic relationship.

GENETICS **p.31** GENETIC MODIFICATION **p.38** EXISTENCE OF GENES **p.106** THE DOUBLE HELIX **p.142** CENTRAL DOGMA OF BIOLOGY **p.172**

# CRISPR Gene Editing Tools

**KEY SCIENTISTS:** FRANCISCO MOJICA • JENNIFER DOUDNA • EMMANUELLE CHARPENTIER

American biochemist Jennifer Doudna (1964–), inventor of the revolutionary gene-editing tool CRISPR, in the Li Ka Shing Center at the University of California, Berkeley, USA.

Genetic engineering is a series of technologies that enable the editing of the genetic code of an organism. It is generally used to introduce a functioning gene from one organism into another, unrelated one. In 2013, the CRISPR technique (pronounced 'crisper') was developed, greatly simplifying this editing process. CRISPR stands for 'clustered regularly interspaced short palindromic repeats'. These are sections of DNA found in the genomes of bacteria that represent the genetic code of viruses that have attacked the cell. The bacteria use an enzyme (Cas9) to splice the viral DNA into its own genetic code as a record so it can launch a fast immune response should the virus attack again. The gene-editing technique harnesses the power of Cas9 to introduce any section of DNA into any part of the genome of any organism. The system works best in single-celled organisms and viruses, and is used to engineer important biochemicals and medical therapies. CRISPR also makes it easier for the eggs or zygote (the first fertilized cell) to create a genetically engineered multicellular organism.

**KEY DEVELOPMENTS**

**CRISPR has democratized genetic engineering to a certain extent, the implications of which are still to become clear because it has given rise to a subculture of biohacking. It is possible to order DNA with a predefined – and even designed – genetic sequence, and biohackers use it to develop products such as brewing yeast that glows in the dark, or more alarmingly injecting people with spurious cures for genetic diseases.**

GENETICS **p.31** GENETIC MODIFICATION **p.38** EXISTENCE OF GENES **p.106** THE DOUBLE HELIX **p.142** CENTRAL DOGMA OF BIOLOGY **p.172**

# Stem Cells

**KEY SCIENTISTS:** ERNEST MCCULLOCH

Mouse neural cells growing in a culture. Neural stem cells can be made to develop into cells found in the central nervous system: neurons, astrocytes and oligodendrocytes.

The human body (in common with other large organisms) is constructed from thousands of cell types. A blood cell is very different from a neuron or a bone cell, for instance, yet all of these types grew from a single first cell, a zygote. As the body develops from a ball of dividing cells into an embryo and on to adult size, the cells it requires differentiate, or specialize, as they divide. Once specialized, a cell can only make copies of itself. However, at the heart of this growth are stem cells, which have the power to grow into any body tissue. Stem cells are now at the cutting edge of medical research. Once the body is fully grown, the number of stem cells at work is drastically reduced, and those that remain are multipotent, meaning they can develop into a limited number of related cell types – red and white blood cells, for example.

Doctors are learning how to reset stem cells back into the pluripotent form, which can differentiate into any cell type. Once that is perfected, stem cells could be used to repair any damaged or diseased body part.

**KEY DEVELOPMENTS**

**One of the techniques used to create stem cells is derived from the technology used in 1996 to make the world's first cloned mammal, Dolly the sheep. The DNA from a skin cell is added to an egg, which then develops into a ball of cells – all pluripotent. In the case of Dolly, the ball was left to develop completely. However, in the case of human cells, the process is halted after a few days so that the stem cells may be harvested before they form a sustainable human embryo.**

CELL THEORY **p.25**

# Clinical Trials

**KEY SCIENTISTS:** JAMES LIND • EDWARD JENNER • GEOFFREY MARSHALL

Once a medical drug has been identified as having a therapeutic effect in an illness, it must be trialled under clinical conditions to see if it is safe. An early example of this kind of trial was carried out in 1747 by James Lind (1716–1794), a doctor in the British Royal Navy. The poor diet on board ships meant that crews suffered from scurvy, at the time a mysterious disease. Lind is reported to have administered six therapies – all acids – to six pairs of sailors. The pair treated with citric fruits recovered, thus leading the way (much later) to linking scurvy with vitamin C deficiency. Today's clinical trials go a step further. The doctors running them treat all candidates the same, testing and monitoring for changes and side-effects. However, only some of the candidates receive the new drug; others get a placebo, or receive an older, already proven remedy. The candidates have no idea which they are given, and importantly neither do the doctors who administer them. This double-blind system, pioneered in the 1940s by Geoffrey Marshall (1887–1982), ensures that no human bias is added to the results.

**KEY DEVELOPMENTS**

**Clinical trials have highlighted the power of the placebo effect, where an inactive substance is given in lieu of a therapeutic agent. This dummy treatment still leads to an improvement in symptoms, such as a reduction in pain or fatigue, although there is no demonstrable impact on an infection or whatever else is causing the disease. A clinical trial must show that a new drug is able to produce an effect significantly greater than a placebo.**

Portrait of James Lind, whose work to eliminate scurvy earned him the byname 'the founder of naval hygiene'.

THE BIRTH OF MEDICINE **p.14** GERM THEORY **p.104** ANTIBIOTICS **p.130**

# Cladistics and Taxonomy

**KEY SCIENTISTS:** ARISTOTLE • CARL LINNAEUS • CARL WOESE

Since its earliest days in Classical Greece, biology has involved classifying life forms into groups that reflect their similarities. At the start of the Scientific Revolution, as the number of species was rapidly increasing, Carl Linnaeus (1707–1778) created a binomial system which assigned two names to every type of life form. This removed any confusion that arose from unregulated common names for organisms. Additionally, Linnaeus organized life into a hierarchy of groupings, or taxons, ranging from kingdom and phylum to genus and species, which clustered organisms according to how similar they were. For example, all mammals are in the Mammalia class, but only the big cats are in the Panthera genus. Linnaeus classified species according to their shared anatomy. A more modern approach, cladistics, is based on evolution. Life is organized by common ancestor, with every member of a taxon having evolved from the same species. The larger the group, the more ancient the members' divergence from this ancestor.

**KEY DEVELOPMENTS**

In the 1990s, Carl Woese (1928–2012) and others proposed adding a new taxon above the level of kingdom. The animal, plant and fungus kingdoms, along with unicellular creatures such as amoebas, all belong to the Eukaryota domain, because they all share a similar cellular anatomy. The rest of life fills two other domains, the Archaea and Bacteria, which incorporate organisms with simpler cell structures.

Illustrations from a 1748 edition of Linnaeus's *Systema Naturae* showing fish classifications. The work was first published in 1735.

NATURAL HISTORY AND BIOLOGY **p.22** DISCOVERY OF MICROORGANISMS **p.64** EXISTENCE OF GENES **p.106** EVOLUTION BY NATURAL SELECTION **p.161**

# Schrödinger's Cat and Other Thought Experiments

**KEY SCIENTISTS:** ERWIN SCHRÖDINGER • GALILEO GALILEI • AL-HAYTHAM ALBERT EINSTEIN • PIERRE-SIMON LAPLACE

Schrödinger's cat is perhaps the most famous thought experiment. Erwin Schrödinger (1887–1961) envisaged it in 1935 as a way of probing contemporary interpretations of quantum mechanics. It entails a hypothetical scenario in which a cat is locked away from view in a box. Within the box is a vial of poison which is linked to a Geiger counter, and set to be released when a radioactive atom decays. The decay is random and unpredictable. Therefore, without opening the box, Schrödinger was unable to say if the poison had killed the cat, so the cat was in a superposition, being both dead and alive at the same time. His point was that this preposterous situation was exactly the one proposed to describe the states of quantum objects.

Thought experiments are powerful tools in science. Galileo, al-Haytham and Einstein were among those who used them. The theory of relativity is said to have begun when Einstein imagined what he would see if he sat on a beam of light. Thought experiments are not meant to offer proof but to guide thinking as scientists develop theories.

The paradox of Schrödinger's hypothetical cat is that it is both alive and dead at the same time.

**KEY DEVELOPMENTS**

Pierre-Simon Laplace (1749–1827) performed a thought experiment now called Laplace's demon. It involves a supernatural intelligence that knows the precise location and motion of every object in the Universe right now, and can use the laws of motion to set out exactly how they will change in every moment of recorded time into the future. Laplace's demon asks whether the Universe is deterministic with a preordained future that was locked in place at the first moment of time.

GREEK PHILOSOPHERS **p.13** THE SCIENTIFIC REVOLUTION **p.18** ACCELERATION UNDER GRAVITY **p.55**
TURING MACHINE **p.138** ATOMIC THEORY **p.159** RELATIVITY **p.163** QUANTUM PHYSICS **p.167**

# Computer Graphics for Modelling

**KEY SCIENTISTS:** EDWARD LORENZ • STANISLAW ULAM • JOHN VON NEUMANN

A woman wearing Virtual Reality glasses.

The value of modelling natural phenomena using software has been recognized since the earliest days of computing, when room-sized primitive computers were used to forecast the weather. In 1963, Edward Lorenz (1917–2008) stumbled across the butterfly effect with his atmospheric model. He found that one tiny change in the model's starting conditions led to wild divergences in results, a feature of mathematics that is at the heart of what is now termed chaos theory. Lorenz's discovery illustrated the dictum 'All models are wrong, but some are useful' – they are not a substitute for collecting real data, but they nevertheless have profound influence.

One familiar use is climate modelling to spot the impact of global warming. Computers can also host virtual chemical and nuclear reactions, so researchers can investigate new drugs and reagents or map the behaviour of subatomic particles. Additionally, a computer makes light work of rendering 3D shapes, which requires calculations using four spatial dimensions.

**KEY DEVELOPMENTS**

One of the first applications of a computer model was in the Manhattan Project that developed nuclear weapons. Nuclear weapons rely on neutrons moving through radioactive material to create a chain reaction. Calculating what conditions were necessary for that stumped human mathematicians. The Monte Carlo method – so named because it uses random chance as in the Monaco casino – was developed to perform the calculations with random changes that resulted in success or failure. This required many iterations to be performed quickly, a perfect job for the Electronic Numerical Integrator and Computer (ENIAC), a US machine introduced in 1946 that was one of the first digital computers.

← ELECTRONICS AND COMPUTATION **p.30** THE INTERNET **p.36** TURING MACHINE **p.138**

# Climate Simulation

**KEY SCIENTISTS:** EUNICE NEWTON FOOTE • SVANTE ARRHENIUS • CHARLES KEELING ROGER REVELLE

The fact that carbon dioxide absorbs heat in greater proportions than more abundant gases in the air was revealed in the 1850s. A century later, decades of meteorological data showed that the climate was warming, and the increase in carbon dioxide and other greenhouse gases produced by human activity was the prime candidate for explaining why. Understanding how human-made emissions were impacting the already complex global climate system became a crucial goal of science. The first climate model from 1956 divided the air into just 500 spaces. Today's models use a 3D grid of 150,000 zones, each with dozens of time-sensitive variables, such as temperature, pressure, cloud cover and humidity. The model allows each cell to exert an influence on the conditions in the surrounding cells as they would in reality each hour of the day. However, the model can map these changes much more quickly than that. The resolution of simulations is governed by processing power, and climate scientists are equipped with some of the world's biggest super computers so they can predict Earth's future atmosphere.

**KEY DEVELOPMENTS**

**The climate is a chaotic system, where small changes at the start can create wildly different outcomes, so each climate simulation seeks to represent the atmosphere in ever more detail. There is often controversy about the veracity of models, and so they are tested by being backcast, or run backwards. This reveals whether they can arrive at the conditions recorded in the past using today's conditions. If so, we can be more confident about what the model says about the future.**

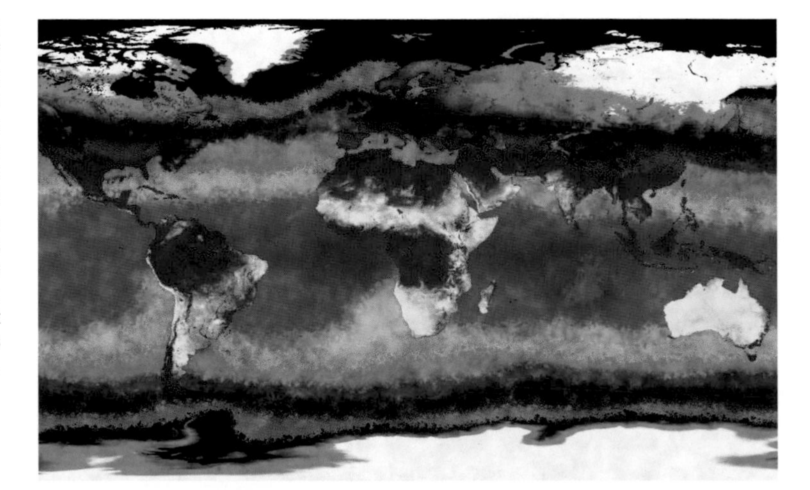

A conceptual illustration of global warming across the entire globe. Earth's average surface temperature has risen by 1.18 degrees Celsius (2.12 degrees Fahrenheit) since the late nineteenth century, and most of that increase has been since 1980.

ELECTRONICS AND COMPUTATION **p.30** ENVIRONMENTAL SCIENCES **p.35**
ANTHROPOGENIC CLIMATE CHANGE **p.178**

# Machine Learning

**KEY SCIENTISTS:** ALAN TURING

One of the features of artificial intelligence (AI) is a computer's ability to programme itself. This process, known as machine learning, is used by AI for developing pattern recognition, such as that used to read text, recognize faces or comprehend speech. Machine learning requires a neural network, a device that emulates the way some parts of the brain work, and can be hardware or software. The network separates a computer input from the output by many layers of interconnected nodes, thus creating a very large number of routes through the network. Machine learning begins with training, where the device learns, for example, to recognize pictures of cats by receiving millions of pictures of cats and other things. At first the code that makes up the pictures travels through the network more or less at random and produces one of two outputs (cat/not cat), also at random. A pathway that produces a true output is favoured over false ones, and eventually, after perhaps millions of trials, the AI is able to recognize a cat picture (or speech, faces, or anything else) just from the path its data take through the network.

**KEY DEVELOPMENTS**

**There are two kinds of AI. The more common is a narrow AI, which is at work in pattern-recognition software in smart speakers or facial identity systems. It can learn a set of tasks better than a human without getting tired or bored, but it does not know that there are things it cannot do. The second kind is a general AI, which matches our own ability to direct our learning to tackle new tasks. A truly general AI is still probably theoretical, although expert systems, where scientists give the AI a database of human knowledge, are certainly making AIs very clever.**

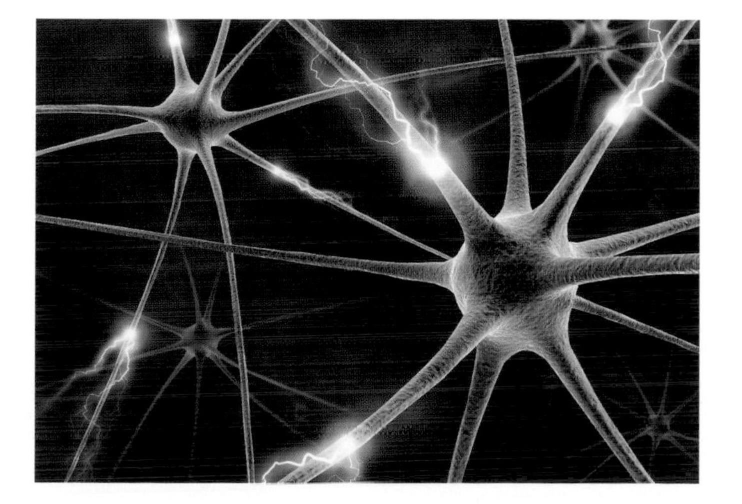

Computer-generated image of a neural network. AI connects and grows in ways that are similar to the workings of the human brain.

ELECTRONICS AND COMPUTATION **p.30** THE INTERNET **p.36** TURING MACHINE **p.138**

# Big Data

**KEY SCIENTISTS:** ALAN TURING • CLAUDE SHANNON • MARK ZUCKERBERG LARRY PAGE • SERGEY BRIN

The collection and analysis of Big Data assist in a vast range of professions and businesses, from medicine to marketing.

The Internet was first envisaged in the 1960s as a way to connect computers. In the 2000s, Web 2.0 brought together many human users – about half of the world's population by 2018 and rising. Now we are at the start of a new phase, the Internet of Things. Already there are three devices connected to the Internet for every human on Earth, and by 2030 more than 50 billion gadgets, from cars and weather stations to fridges and heart monitors, will be sending and receiving data. The concept of Big Data is to collect all this disparate information and search for patterns that might reveal unseen links. This method of analysis could be a game-changer for the study of natural, planet-wide phenomena, which are currently limited by the amount of data available. Big Data could also be used to train AI, creating a 'system of systems' that orchestrates services and utilities – water and energy supplies, transport, healthcare and communications – thus making them all responsive to each other and better able to work optimally.

**KEY DEVELOPMENTS**

**Big Data can be useful in evidence-based medicine where measurable signals are used in the diagnosis of illness and trigger treatments only when they are known to be effective. It is standard medical practice to keep detailed records of each patient, and anonymized versions of the data could be combined to search for early signals of developing disease, or link the successes or failures of particular treatments with previously hidden factors.**

ELECTRONICS AND COMPUTATION **p.30** THE INTERNET **p.36** TURING MACHINE **p.138**

# Planetary Rover

**KEY SCIENTISTS:** EUGENE SHOEMAKER

While the first decade of the Space Race was focused on putting humans into space and sending them to alien worlds, by the 1970s attention had largely shifted to the cheaper and safer alternative of sending mobile, wheeled robots. The first of these rovers were Lunokhod 1 and 2, sent by the Soviet Union to the Moon, where they spent a combined 15 months on the surface and covered 39 kilometres (24 miles) of ground. Their modern successors are the NASA Mars rovers, Spirit, Opportunity and Curiosity. Together, they have racked up nearly 30 rover years working on the Red Planet. Only Curiosity is still functioning. A Mars rover is equipped with stereoscopic cameras that create 3D views of the planet's surface. These are used by controllers back on Earth to plot routes, while the rover maintains autonomy over the speed and power used to tackle obstacles. The rovers are equipped to analyze rocks and minerals, searching for signs of water and life. They do this by collecting samples with scrapers and scoops, which are analyzed by laser spectroscopes and on-board laboratories.

**KEY DEVELOPMENTS**

In 2020, Curiosity was the only active rover on Mars. It was joined in 2021 by NASA's rover Perseverance, and Ingenuity, a little dual-rotor drone for scouting territory. A European rover, Rosalind Franklin, is also slated to arrive in the 2020s. In 2018, China landed Chang'e 4, the first rover to reach the far side of the Moon, and in 2020 the Chang'e 5 mission sent back the first samples of rock from there.

This artist's impression depicts NASA's Mars 2020 rover studying a rock outcrop on the surface of the planet.

THE SPACE RACE **p.32** EXOPLANETS **p.148** ORIGIN OF THE SOLAR SYSTEM **p.179**

# Index

Illustrations are indicated in *italic*
Main entries are indicated in **bold**

**D**

### E



### F



### G

# Picture Credits

**10** Wellcome Collection CC **12** PRISMA ARCHIVO/Alamy **13** Wikimedia Commons **14** Wikimedia Commons **15** Metropolitan Museum of Art, New York/Harris Brisbane Dick Fund, 1926 **16** Wellcome Collection CC **17** Wikimedia Commons **18** Wikimedia Commons **19** World History Archive/Alamy **20** Science History Images/Alamy **21** Ivy Close Images/Alamy **23** North Wind Picture Archives/Alamy **24** Wellcome Collection CC **25** Library Book Collection/Alamy **26** Wellcome Collection Public Domain **27** Wikimedia Commons **28** Granger Historical Picture Archive/Alamy **29** Science History Images/Alamy Stock **30** Science History Images/Alamy Stock **31** Wikimedia Commons **32** NASA **33** Dietmar Temps/Alamy **34** Keystone/Hulton Archive/Getty Images **35** Avalon/Bruce Coleman Inc/Alamy **36** Creative Commons **37** Vassar College Library, Archives and Special Collection **38** Wikimedia Commons **39** Jbourijai **40** Rémih/Wikimedia Commons **42–3** Photo 12/Alamy **445** WENN Rights Ltd/Alamy **46–7** Wellcome Collection CC **48–9** Stefano Politi Markovina/Alamy **51** Wellcome Collection Public Domain **53** Wikimedia Commons **54** Science History Images/Alamy **55** The Print Collector/Alamy **56–7** Science History Images/Alamy **58–9** ipsumpix/Corbis **61** Science History Images/Alamy **62–3** Photograph by Mike Peel (www.mikepeel.net)/Creative Commons **65** World History Archive/Alamy **67** The Granger Collection/Alamy **69** Wellcome Collection CC **71** Wellcome Collection CC **72–3** Wellcome Collection Public Domain **75** Wellcome Collection CC **77** Metropolitan Museum of Art, New York. Purchase, Mr. and Mrs. Charles Wrightsman Gift, in honor of Everett Fahy, 1977 **78–9** Wellcome Collection Public Domain **80–1** Wellcome Collection CC **82–3** Chronicle/Alamy **84–5** World History Archive/Alamy **86–7** Science History Images/Alamy **88** The Granger Collection/Alamy **89** Wikimedia Commons **91** Wellcome Collection CC **93** INTERFOTO/Alamy **95** The Reading Room/Alamy **97** Granger Historical Picture Archive/Alamy **98–9** World History Archive/Alamy **101** Rémih/Wikimedia Commons **102–3** Science History Images/Alamy **105** Wikimedia Commons **107** Natural History Museum/Alamy **108** Reading Room 2020/Alamy **111** Science History Images/Alamy **113** Wellcome Collection CC **115** FLHC10/Alamy **116–17** Granger Historical Picture Archive/Alamy **119** Bryn Mawr Special Collections **120–1** Granger Historical Picture Archive/Alamy **123** Universal Images Group North America LLC/Alamy **125** The Picture Art Collection/Alamy **126–7** Wikimedia Commons **129** Science Museum **131** Wikimedia Commons **133** Pictorial Press Ltd/Alamy **135** Smithsonian Institution CC **136–7** United States Department of Energy/Wikimedia Commons **138** Schadel/Wikimedia Commons **139** Central Historic Books/Alamy **140–1** Roger Ressmeyer/Corbis/VCG via Getty Images **143** A Barrington Brown, © Gonville & Caius College/Coloured by Science Photo Library **145** Fred the Oyster/Wikimedia Commons **146–7** Science History Images/Alamy **149** NASA/Troy Cryder **150–1** NASA **152–3** Stocktrek Images, Inc /Alamy **154** NASA/JPL/California Institute of Technology **156** Wellcome Collection CC **157** Wellcome Collection CC **158** Wikimedia Commons **159** Wellcome Collection CC **160** Wellcome Collection CC **161** World History Archive/Alamy **162** Science History Images/Alamy **163** Wikimedia Commons **164** Granger Historical Picture Archive/Alamy **165** incamerastock/Alamy **166** Interfoto/Alamy **167** Science History Images/Alamy **168** World History Archive/Alamy **169** GL Archive/Alamy **170** NASA/JPL-Caltech/Univ. of Ariz./STScl/CXC/SAO **171** domdomegg/Wikimedia Commons **172** BSIP SA/Alamy **173** Nancy R. Schiff/Getty Images **174** Shutterstock **175** NASA/JPL/California Institute of Technology **176** NASA/Dana Berry **177** GiroScience/Alamy **178** Pictorial Press Ltd/Alamy **179** RGB Ventures/SuperStock/Alamy **180** CERN **182** GL Archive/Alamy **183** Wellcome Collection CC **184** The History Collection/Alamy **185** National Institute of Standards and Technology/Wikimedia Commons **186** North Wind Picture Archives/Alamy **187** Wellcome Collection/Science Museum, London CC **188** Wellcome Collection CC **189** Xinhua/Alamy **190** Wellcome Collection CC **191** Wellcome Collection CC **192** Metropolitan Museum of Art, New York, Harris Brisbane Dick Fund, 1936 **193** Wellcome Collection CC **194** Wellcome Collection CC **195** Richard Wainscoat/Alamy **196** Allan Swart/Alamy **197** James King-Holmes/Alamy **198** Jim West/Alamy **199** Department of Energy/Wikimedia Commons **200** CERN **201** Lawrence Berkeley National Lab, Roy Kaltschmidt, photographer **202** Science History Images/Alamy **203** Sciencephotos/Alamy **204** Science History Images/Alamy **205** Wellcome Collection/Alec Jeffreys CC **206** The Washington Post/Getty Images **207** Wellcome Collection/Yirui Sun CC **208** Wellcome Collection CC **209** Wellcome Collection Public Domain **210** Mopic/Alamy **211** Alamy **212** Boscorelli/Alamy **213** Science Photo Library/Alamy **215** NASA/JPL-Caltech